SECOND EDITION

AUSTRALIAN MAGPIE

Biology and Behaviour of an Unusual Songbird

GISELA KAPLAN

© Gisela Kaplan 2019

All rights reserved. Except under the conditions described in the *Australian Copyright Act 1968* and subsequent amendments, no part of this publication may be reproduced, stored in a retrieval system or transmitted in any form or by any means, electronic, mechanical, photocopying, recording, duplicating or otherwise, without the prior permission of the copyright owner. Contact CSIRO Publishing for all permission requests.

A catalogue record for this book is available from the National Library of Australia.

ISBN: 9781486307241 (pbk.)
ISBN: 9781486307258 (epdf)
ISBN: 9781486307265 (epub)

Published by:

CSIRO Publishing
Locked Bag 10
Clayton South VIC 3169
Australia

Telephone: +61 3 9545 8400
Email: publishing.sales@csiro.au
Website: www.publish.csiro.au

Front cover: Australian Magpie, photo courtesy of Andrew Skeoch, www.listeningearth.com.au

Photographs are by the author unless stated otherwise.

Set in 11/13.5 Adobe Minion Pro & Helvetica Neue LT Std
Edited by Anne Findlay, Princes Hill, Melbourne
Cover design by James Kelly
Typeset by Desktop Concepts Pty Ltd, Melbourne
Index by Bruce Gillespie
Printed in China by Toppan Leefung Printing Limited

CSIRO Publishing publishes and distributes scientific, technical and health science books, magazines and journals from Australia to a worldwide audience and conducts these activities autonomously from the research activities of the Commonwealth Scientific and Industrial Research Organisation (CSIRO). The views expressed in this publication are those of the author(s) and do not necessarily represent those of, and should not be attributed to, the publisher or CSIRO. The copyright owner shall not be liable for technical or other errors or omissions contained herein. The reader/user accepts all risks and responsibility for losses, damages, costs and other consequences resulting directly or indirectly from using this information.

Acknowledgement
CSIRO acknowledges the Traditional Owners of the lands that we live and work on across Australia and pays its respect to Elders past and present. CSIRO recognises that Aboriginal and Torres Strait Islander peoples have made and will continue to make extraordinary contributions to all aspects of Australian life including culture, economy and science.

The paper this book is printed on is in accordance with the standards of the Forest Stewardship Council®. The FSC® promotes environmentally responsible, socially beneficial and economically viable management of the world's forests.

Contents

	Acknowledgements	iv
	Introduction	v
Chapter 1	Origins	1
Chapter 2	Which is the 'real' magpie?	15
Chapter 3	Anatomy	27
Chapter 4	The brain and the senses	41
Chapter 5	Diet and cognition in foraging	57
Chapter 6	Managing a territory	79
Chapter 7	Bonding and breeding	99
Chapter 8	Caring for the young	115
Chapter 9	Social rules and daily life	139
Chapter 10	Song production and vocal development	159
Chapter 11	Communication	179
Chapter 12	Magpies and humans	199
	Epilogue: The success of magpies	213
	Plates	215
	References	223
	Index	255

Acknowledgements

A book of this kind would not be possible without the substantial endeavours by researchers and by committed ornithologists. I also wish to extend a general thank-you to the members of the public from most states and territories of Australia who, over the years, have taken the trouble to write to me about their experiences with magpies, and even sent tape recordings. This information has given me a valuable insight into the magpie's role in Australia's culture and psyche, and alerted me to behaviours that may have been rare or unusual and required explanation. I wish to thank the New England Branch of WIRES for its permission to allow access to some of its local rescue records, and Rebecca Con for preparing WIRES data. A special thanks also to the University of New England Natural History Museum for allowing me free use of its selection of magpie specimens to study anatomical details. The museum is an invaluable resource. My sincere thanks go to Professor Lesley Rogers who has read the entire manuscript of this edition closely and has made many valuable suggestions. My thanks go also to the University of New England for supporting my grant applications, which have allowed me to conduct ongoing field research in magpie behaviour; to Nick Alexander (first edition) and Briana Melideo (second edition), of CSIRO Publishing, and the entire staff at CSIRO for their guidance and conviction in the importance of this project.

Introduction

In our bird-rich nation, the Australian magpie, one of the foremost songbirds in the world, is arguably the most researched and also perhaps the best known native bird. According to the survey conducted by BirdLife Australia in conjunction with *Guardian Australia* in December 2017, it is now 'official' that the magpie is the most popular bird in Australia.

Indeed, the Australian magpie enjoys the status of a culturally important icon. The kookaburra may signify Australia but the magpie has a special place in the hearts of Australians. This may be so because magpies, unlike kookaburras, are found almost anywhere in this country. They often share suburban backyards and rural properties with human populations.[1] Because magpies are territorial, they tend to stay in the one place for as long as they can hold it and this tends to facilitate contact with humans.

Added to the tremendous general interest in the species, a significant number of people form strong personal attachments to magpies. Sometimes, of course, magpies cause us problems and these too are part of our heritage and interaction, as will be discussed.[2] We can get to know magpies well because they also tend to live relatively long lives. Their life expectancy is around 25 years, and some claim even longer, up to 30 years.[3] A magpie's life span is thus greater than that of most domestic companion animals and such a stable presence may also bring about close acquaintances with long-term human residents. Most importantly, though, magpies themselves show signs of being amenable to warming to humans as companions (see figure on next page).

For some, it is irresistibly attractive that magpies volunteer to come to the back porch and actually communicate with the inhabitants of the house. Some may even stray into the house on foot and, without any sign of fear, investigate the kitchen and living quarters. Among the nicest stories are those recounting the transfer of the relationship between bird and human to the magpie's offspring, thus increasing the number of magpies appearing at the back porch. Friendships with magpies will be discussed in great detail in the last chapter, including why

Magpies take an active interest in their environment and there is little that they seem to miss.

such friendships are possible at all and why the word 'friendship' applies in its inherent meaning of reciprocity.

Part of the attraction is that these birds remain free ranging and self-sufficient and yet choose to befriend humans. Their demeanour is often so expressive that people are fascinated and feel they have personal access to the magpies and to their lives, as if being handed a small looking-glass into the natural world which, in turn, often leads to spending more time watching their behaviour.

It is not uncommon that those who claim to have no particular knowledge of birds often become very good naturalists and ethologists in the process of these evolving friendships with magpies. There may be some unusual magpie behaviour that this book cannot readily explain, especially when the observed behaviour is rare and concerns very specific moments. However, this does not invalidate anecdotal observations. On the contrary, over the last century, the many very active ornithological clubs and naturalists in Australia have been studious in recording their observations in print. Thus, details of magpie behaviour have accumulated and substantially contributed to maintaining and spawning research interest in magpies.

The fields of both ethology/animal behaviour and ecology have been particularly fruitful in recent years and as a result following innovative research attitudes to animals, and especially to birds, these fields have also changed substantially. We now know that birds share not only a range of emotions common to humans, but they have families, suffer when a member is lost, and, most surprisingly, have been found to think, solve problems, have phenomenal memories, make decisions and will barrack for their partner and fiercely defend their young. The astonishing advances that we have made in understanding what birds are about have been infinitely strengthened by cross-disciplinary research in neuroscience, ethology, comparative psychology, ecology and biology. It is doubtful whether any of these advances would have been possible without such cross-disciplinary input.

When the first edition of this book came out, it was based on 10 years of my research on magpies. This second edition has been written 15 years later. Because magpies offer so many surprises in their behaviour and there has been so much commitment to find out more about this species, those 15 intervening years have been marked by ongoing research including up to a quarter of a century of my own work specifically on magpies. In these years we have learned that birds are a good deal cleverer than we ever thought, have abilities far exceeding anyone's expectations and, moreover, have the added advantage that their ability to learn vocalisations and their willingness to raise offspring in pairs or groups make them eminently suited for comparative work on the evolution of language and cognition.

This book is meant to be accessible to everybody and transparent in what it wants to convey. It can be read vertically and horizontally. Horizontal reading is from one end to the other as it is written and spiked with many illustrations to exemplify points. Vertical reading offers the opportunity to follow up on the scientific background that is incorporated and literally covers most of the research ever published about magpies, even if summarised briefly, and can be pursued in whatever direction a reader's interest is piqued. It is not a matter of immodesty that my own research is included in this book but a reflection of how much more we know now than we did a decade ago. The book also includes many of my own anecdotal observations for which we do not as yet have full explanations – their value is that they were observed in the field and may lead to further enquiries.

In the first edition, I expressed the hope that research on all aspects of magpie biology and behaviour will continue to thrive. Now, 15 years later, it can be confirmed that, indeed, research on magpies has been thriving. My own research on cognitive behaviour, song production and communication since the publication of the first edition in 2004, has discovered remarkable traits and abilities in Australian birds generally (cf. also Kaplan[4]) and in magpies specifically, findings that will be presented at least in summarised form in this book. Biogeography, palaeontology and taxonomy of Australian birds have taken a very dramatic turn since 2004.

Indeed, there is now international recognition that all modern songbirds first arose in Australia because some lineages survived the mass extinctions and survived only in East Gondwana. So dramatic are some of the discoveries that it has changed our thinking and expanded our knowledge about this continent, about its bird life and other vertebrates, even about species interactions and internal dispersal and speciation. It is hard to overstate the magnitude of the new understanding of our island world as an 'Out of Australia' narrative for birds, their natural history and species evolution since the mass extinction 65 million years ago.

This new edition aims to explore all these new facets of our current knowledge of the Australian continent and its avian species – with the magpie taking centre stage again – and peer into its daily life, its brain and abilities and its problems and successes. Thanks to the substantial amount of research and field work that has been undertaken on magpies alone over the last decade or so, we can now celebrate the profile of this bird as perhaps we can no other.

Therefore, this book is not just a new edition but is substantially based on new material that was not available before 2004, when the writing of the first edition finished. Indeed, the years since the publication of the first edition have seen such major advances and insights into our thinking about birds, in terms of origin, cognition, physiology and climate, that this is, apart from my personal partiality to magpies, compelling reason enough to have written it.

This book aims to make this special Australian more accessible to the many people who have an abiding interest in magpies. Research articles provide a very useful information service to scientists but tend to be less accessible to the general public and the bulk of knowledge rarely trickles through to a wider audience other than in the occasional flash headline. And yet, when it comes to native wildlife I feel that there continues to be an urgent need to bridge this gap because everybody should be able to have access to the latest insights and knowledge on our wildlife without having to make very special efforts to find and decipher complex scientific papers.

Increasingly, dictionaries, handbooks and research publications will expand our knowledge of Australian native fauna but rarely is one given the luxury to write a book-length work about one native bird species. We still have a good deal to do to make Australian fauna better known. I can think of a no more promising and enjoyable way of doing so than with the Australian magpie.

Gisela Kaplan
February 2018

Endnotes

[1] Blakers *et al.* 1984
[2] Jones 2002
[3] QNPWS 1993
[4] Kaplan 2015

Introduction | ix

A magpie group's walk to the back porch for titbits is a familiar experience to many Australians who find magpies irresistible.

1

Origins

In 2013, a small online article in the *Sydney Morning Herald* related an enchanting story about the Australian magpie in Aboriginal Dreaming and the creation of daylight. As retold by Hancock[1] and Rule[2], it was the clever magpie that raised the sky and helped emus to straighten their necks, kangaroos to hop and wombats to leave their burrows because magpies had decided to hold up the sky.

According to the Dreaming, the sky was once so close to the ground that trees could not grow, people had to crawl and all the birds were forced to walk everywhere. The days were dark and cold and one day the magpies decided to hold a meeting on how to end this undesirable situation. They settled on a solution. Working together they managed to prop up the sky with sticks, but it threatened to break the sticks and collapse to Earth again – so they went to higher hills, took a long stick in their beaks and kept pushing the sky higher and higher, until they reached the highest mountain in the whole land. Then, with a special heave, they gave the sky one last push! The sky shot up into the air, and as it rose it split open and a huge flood of warmth and light poured through on to the land below. The animals wondered at the light and warmth, but more at the incredible brightly painted beauty of the Sun-Woman. The whole sky was awash with beautiful reds and yellows. It was the first sunrise.

So that is why, according to this Dreaming, every morning when the Sun-Woman wakes and lights her early morning fire, to this day all the magpies greet her with their beautiful song, perhaps also to remind everyone else of their important role in holding up the sky. The magpie's song is reflected in its Noongar

Era	Period	Epoch	Age
Cenozoic	Quaternary	Anthropocene	present
		Holocene	0.01 mya
		Pleistocene	1.8 mya
	Tertiary	Pliocene	5 mya
		Miocene	24 mya
		Oligocene	34 mya
		Eocene	55 mya
		Paleocene	65 mya
Mesozoic	Cretaceous	Late	99 mya
		Early	144 mya
	Jurassic	Late	159 mya
		Middle	180 mya
		Early	206 mya

Fig. 1.1. Ages of the Earth. The Jurassic period (the age of dinosaurs) is the best-known period. The Jurassic period extends from 206 million years ago to 144 mya. The Cretaceous begins 144 and extends to 65 mya when the mass extinction occurred. Latest evidence suggests that at least some of the bird lineages that survived the mass extinction first evolved in the early Cretaceous (meaning that they are older than those arising in the late Cretaceous). The K-T boundary of 65 mya indicates the mass extinction. In the text, reference is usually made to the periods rather than to dates. Most songbirds we know today already existed by the end of the Tertiary period but many birds have gone extinct over the last 100 000 years (during the Holocene Epoch) and, at an accelerated rate, in the newly identified Epoch of the Anthropocene.[3]

name *koolbardie*. The mining town of Coolgardie, Western Australia, means 'magpie' in the Goldfields' Aboriginal dialect.

The story is one of many different ones concerned with creation. Notably, magpies play a central and positive role in it, suggesting that magpies have the determination and cognitive ability to resolve problems as we now know, scientifically, they do. This tale (and there are many more about magpies) also confirms the very long-standing relationship humans have had with magpies on this continent.

One might use this tale also as a fitting symbol for the end of the devastating events in geological history: the mass extinction events of ~65 million years ago (mya) (see Fig. 1.1 for detail of the ages of the earth). This particular mass extinction event (there were several before) is thought to have been flood basalt eruptions combined with dramatic falls of sea levels and was followed by a comet (or asteroid) that struck the sea bed near the Yucatan Peninsula in the now Gulf of Mexico ~65/6 mya.

By generating a vapour-rich impact plume, vast amounts of dust particles plunged the world into darkness and cold, inhibiting photosynthesis for many years. It is thought that the consequences of lingering dust completely occluded sunlight for up to six months, thus seriously disrupting continental and marine food chains. This then resulted in the death of most plant life and phytoplankton, which would also kill many of the organisms that depended on them. When the sunlight returned, the Earth had lost 75% of all flora and fauna. The most popularised species among the dramatic losses are the dinosaurs because they left plenty of fossil evidence around the world. Over time, many dinosaur species were also recovered from Victoria (Dinosaur Cove), Queensland (inland Winton, Riversleigh in the north), South Australia (Coober Pedy) and Western Australia (Broome) – testament to their demise also on Australian soil.

Among the 25% of flora and fauna that survived there were also some birds but for a long time it was thought that these had derived from the Northern Hemisphere, somewhere on the Eurasian landmass, seemingly supported by early fossil finds of feathered dinosaurs in Europe and in China.

The lineages of birds that survived did so in East Gondwana, i.e. in Australia.[4] Not only is Australia the cradle of all true songbirds[5] but, as is slowly being established and confirmed, the cradle of many other avian clades and families as well.[6] Diversification of class aves thus seems to have begun during the Cretaceous period.[7] The oldest evidence of avian presence in Australia consists of 105 million-year-old fossilised footprints and feather deposits found at Dinosaur Cove on the coast of southern Victoria. The researchers thought that the tracks belonged to a species about the size of a great egret or a small heron.[8] It has been suggested that the mass extinction at the end of the Cretaceous period led to a period of recovery, followed by 'explosive' speciation.

About 200 mya, Gondwana was a compact supercontinent but eventually broke up over a 150-million-year period (Fig. 1.2). After that, Australia was on its own drifting northwards with a slight counterclockwise direction edging the continent into its current position.

The final confirmation of Australia as the origin of songbirds[4,5,9,10,11] has led to many excellent studies and findings but also, more than ever, put Australian birds in a fresh light – native species and subspecies are still being identified and reassigned to this day. It also had to be asked how and when species left Australia to populate other continents. Such questions required a wider context than a within-Australia inquiry.

While this work, at times, may seem distant and abstract to the uninitiated, it has many vast implications, not just to unravel how the evolution of birds actually worked and in which order species evolved but also for the management of species alive today. It may mean, for example, that a species, suddenly declared to fall into several subspecies, i.e. numerically subdivided, may reveal one of its subspecies to be critically underrepresented and thus endangered. This new information can then lead to new conservation management plans. Equally, it is important to know what affects diversification and decline.

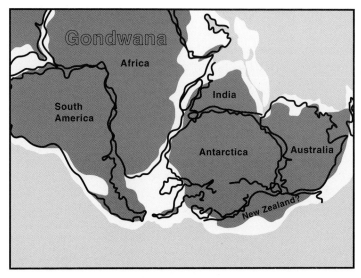

Fig. 1.2. Gondwana ~200 mya consisting of many landmasses that now form separate continents or islands but also of those such as much of S and SE Asia (e.g. India, Thailand, Malaya) that floated northwards and now belong to other continents or subcontinents. New Zealand is marked with a question mark because, although known to have been part of Gondwana, for geological reasons has recently been considered as a mini continent. Africa, India and parts of the now Middle East separated first while the link between South America, Antarctica and Australia endured much longer.

This state of flux has affected the magpie as well and in some quite unexpected ways. Since the Australo-Papuan origin of songbirds was internationally confirmed, relationships involving the magpie have been changed several times and in somewhat incompatible ways. To explain this, we can take one brief look at how songbirds may be subdivided (Fig. 1.3). Importantly, there are two superfamilies (to which not all taxonomists subscribe) separating the Passeridae (small songbirds) from corvid-like species, the Corvoidea as shall be explained.

Fig. 1.3 shows only one possible model. When Cracraft and colleagues proposed an avian tree of life, they constructed the phylogenetic relationships among modern birds in such a way that the magpie ended up in a different and unexpected family.[6] They classed magpies under a superfamily of bush-shrikes (Malaconotoidea) – No. 29 in Fig. 1.3 below, where it is just listed as a family within the superfamily of Corvoidea. This superfamily contains the bushshrikes (Malaconotidae), helmetshrikes (Prionopidae), ioras (Aegithinidae), vangas (Vangidae) and the Australian butcherbirds, magpies, currawongs and woodswallows (Artamidae). Manegold[12] even saw phylogenetic affinities between the vangas (No. 22 in Fig. 1.3) and magpies and allies, as well as drongos (Dicruridae) and monarchs (Monarchidae). Vangas (Vangidae) occur in Madagascar and also in mainland Africa.

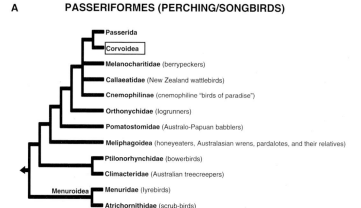

Fig. 1.3. The superfamily of Corvoidea and the families that belong to corvid-like species. A, The various branches of songbirds (simplified). There are over 10 000 species of songbirds in the world. Corvoidea alone make up over 800 species. B, The list of Corvoidea shows the number of families usually counted in this group. The first 15 are families found in Australia. The lower 11 are part of a radiation that originated in Australia.[13]

Overseas connections

What is remarkable about these attributions is that the Australian magpies and allies are grouped together with a vast diversity of shrike-like songbirds widespread in Africa. Norman and colleagues found magpies and allies are even more closely related to the African bush-shrikes and allies (Malaconotidae) than to the Australo-Papuan-centred fantails (Rhipiduridae), birds of paradise (Paradisaeidae) and monarchs (Monarchidae).[14]

In 2012, Jerome Fuchs' analysis, using both mitochondrial and nuclear DNA, shed some further light on this puzzling arrangement by suggesting that this superfamily originated in Australasia, and that some ancestors made the probably substantial sea crossing from Australia to Africa some 33–45 mya (the late Eocene) where they then further diversified.[15]

Manegold[12] thought that the radiation into Africa may have been undertaken by the last common ancestor (stem species) of Artamidae, Cracticidae and Vangidae and he imagined such a bird to have looked something like this:

'could have been a medium-sized, stoutly built passerine with quite long and pointed wings, a stout, slightly decurved bill with a wide gape and a characteristic colour pattern, i.e. slaty bluish-grey with a blackish tip. In addition, sexual monomorphism as well as a predominantly black, white and grey plumage colouration.'

While, from a modern perspective, this trans-ocean migration may sound almost improbable, it is important to remember that neither Australia nor the rest of the world were biogeographically the same as today.

In order to assess how some radiations of birds between countries and continents might have happened, it is important to try and arrive at some plausible narrative. For instance, the question is why do avian species living in Madagascar show a definite relatedness to magpies and how could a radiation from Australia across the Indian Ocean have been possible at all? Seabeds were not flat but consisted of volcanoes, ridges and plateaus that dynamically could be subducted, could subside or erupt and, depending on sea levels, were either submerged or, sometimes, violently, forced above sea level. We know this well of Hawaii and New Zealand. We also know that low sea levels created temporary land bridges of the Javan Trough (see Fig. 1.5B) or between New Guinea and Australia, but the geological history of the Indian Ocean is better known to oil companies and mariners than to biologists or bird lovers.

From today's perspective, it is hard to imagine that the Indian Ocean had to be created by cracks in the Australian, African and Antarctic plates. As Fig. 1.4 shows, it took well over 50 million years before the waterways off western Australia could be called an ocean.

Still, the problem of transfer of species from one continent to another across a substantial ocean remained problematic although imaginable. This is a problem that Schwarz and colleagues faced – how to explain how African bees got to

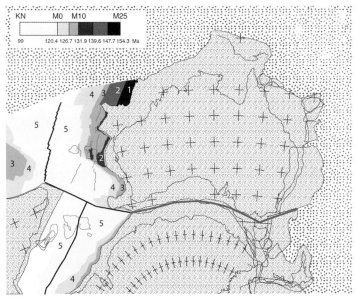

Fig. 1.4. In the beginning of the Cretaceous period (144 mya) there was no Indian Ocean yet. The rift had only just occurred and the ocean took millions of years to 'grow' to its current dimensions and to spread a seabed around Australia. Seabeds around western Australia: current outline of Australia and Antarctica drawn in. Numbers indicate the expansion of seabeds to the west of Australia and around Antarctica. Numbers (1–5 from earliest to latest) refer to periods of water/seabed formations 1: 154.3 mya; 2: 136 mya; 3- 4- 5: only 99 mya sea floors from east of Africa, from the Antarctic and Western Australian finally meet. (Adapted from Müller *et al.*)[16]

Australia in a prehistoric/pre-human single colonisation event.[17] They puzzled how bees could have made their way from southern Africa to western Australia millions of years ago and came up with the suggestion that there must have been islands they were able to use. Indeed, they considered the largely submerged elements of the Kerguelen Plateau and Broken Ridge provinces, as a possible explanation (see Fig. 1.5A).

The Kerguelen Islands, ~3000 km to the south-west of Australia, including also the Heard and McDonald Islands, form just small specks in a vast ocean today. However, these are only the visible part of a submerged mini continent that was once three times the size of Japan. Researchers found soil layers in the submerged land indicating that indeed the plateau was once above sea level, at least for three periods between 100 mya and 20 mya.[18] Researchers like Duncan[19] even concluded that a chain of islands, including the mini continent and the Broken Ridge complex, might have formed an almost unbroken bridge between Africa and Australia and more continuously over the best part of 100 million years in exactly the time-frame of interest to us in terms of bird colonisation events of Africa from

Australia. Explanations of geological events open opportunities to also explain avian dispersal events. In any case, these periods fall well within possible periods of radiation both for birds and bees.

As Fig. 1.5B shows very clearly, the ocean floor of the Indian Ocean has quite a number of submerged ridges, mountains and plateaus. For instance, take the Broken Ridge to the north of the Kerguelen mini continent that was torn from the Kerguelen Plateau because the latter attached to the Antarctic plate while the former attached to the greater Indian plate moving northwards. Over long periods of time they drifted well apart.

This remarkable circumstantial evidence in reconstructing possible flight paths suggests at least that if a bee can get from Africa to Australia (Fig. 1.5A), the journey could plausibly also be made by a bird and could also be undertaken in reverse order (from Australia to Africa), as long as such trans-ocean travel coincided with the times when these islands were in fact above sea level.

Since Fuchs *et al.*[15] estimated that an ancestor of the magpie might have arrived in South Africa between 45–35 mya, geological conditions ought to have been favourable for such a venture. Moreover, Frey and colleagues[20] found that at least in the mid-Cretaceous but possibly later, the islands were covered by dense conifer forests, hence were not barren rock formations but places that were habitable and likely to provide food sources.

This means that groups of magpie ancestors could have taken this route, never to return and the island chains all but vanished, but the link between magpies and African bush-shrikes and even of the vangas of Madagascar and on the African mainland is embedded in their respective DNA.[11,14]

The New Guinea connection

Looking for molecular relationships in avian species to the north-east of Australia, namely to Papua-New Guinea, one might imagine would require far less detective work than any cross-ocean radiation to the west. But since the publication of the first edition of this book in 2004, a good deal of exciting molecular and phylogenetic work has been done, and new light has been shed on Australia's avian past. These include a few surprises as to the classification of the Australian magpie.

It has been known for a long time that Australia and Papua-New Guinea are part of one plate, called Sahul (Fig. 1.6B). While sea levels have fluctuated wildly over millions of years (Fig. 1.6A), more recent glacial and interglacial periods have had the most notable effects on Australian coastlines and on the link to New Guinea. At times of glacial periods, sea levels dropped substantially, and for long periods of time, such drops provided exposure of the entire landmass between Australia and New Guinea and also of some areas to the west of New Guinea.

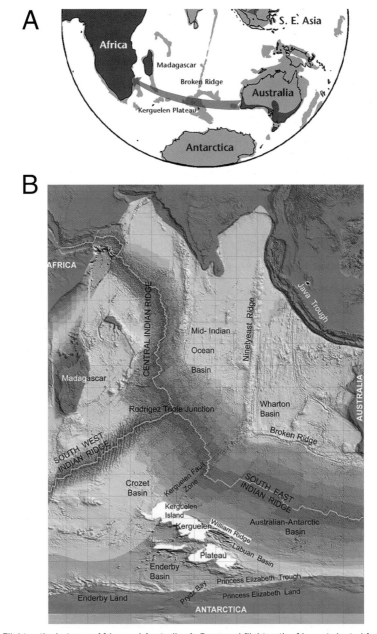

Fig. 1.5. Flight paths between Africa and Australia. A, Proposed flight path of bees (adapted from Schwarz et al.).[17] Note that in prehistoric times the islands of the Kerguelen Plateau and Broken Ridge were close to each other and formed a chain that reached from Africa to Madagascar and to Australia, making ocean crossing possible for winged animals (such as birds and bees). B, The same region of the Indian Ocean but showing its modern underwater landscape, ridges and plateau. The white area indicates a partial above water (sub-aerial) section of the plateau and the well-sculpted areas indicate underwater elevations. Continents of Australia (far right) and Africa (far left) are just visible. Note that the Broken Ridge, once close to the Kerguelen Plateau, has drifted substantially northwards. Ridges are identified and marked as lines (excerpted from Bénard et al.).[21]

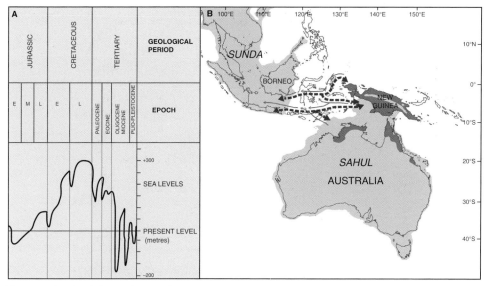

Fig. 1.6. Sea level changes and land exposure. A, Approximate sea level changes from the Jurassic period to the present (reading downwards; E-early, M-middle, L-late), with indications of current sea level; B, Low sea level exposed the landmass, termed Sahul, that included Australia and all its northern islands and Papua New Guinea and far extended the coastline towards modern Indonesia. The dark shading in B refers to the range of the black butcherbird and the arrows (both ways from and to Sunda) suggesting that, at certain times in the past, moving in and out of Sahul/Australia often required no more than island hopping.

To this day, Australia and New Guinea share several species, mammals and birds alike: among birds, for instance, are the southern cassowary, the marbled frogmouth, the palm cockatoo, eclectus parrot and many others.

However, the latest work has again brought some surprises. One was the discovery of fossils in Riversleigh World Heritage Area, north-western Queensland, which were identified to be another species of cracticid, geologically the oldest and the first Tertiary fossil of a cracticid found in Australia.[22] The fossil was named after the describer's father but, as a special touch she added a northern Queensland Aboriginal word *kurrartapu* meaning 'magpie'. The species is now called *Kurrartapu johnnguyeni*. Finding a new genus and species of cracticid from an Early Miocene deposit (Tertiary period) fits in well with molecular estimates for the timing of the cracticid radiation and is thus another important piece of evidence in early passerine evolution.

There are other puzzles that have emerged, namely that specific groups of magpies in the Kimberley and Pilbara are more closely related to eastern than to other local populations. The New Guinean subspecies of the Australian magpie, *Gymnorhina tibicen papuana*, also shares morphological traits with subspecies from north-western and south-western Australia, *Gymnorhina tibicen longirostris* and *G. t. dorsalis*, respectively.[23] Yet the black-backed butcherbird of Cape York, so near New Guinea, is only distantly related to the New Guinean species.

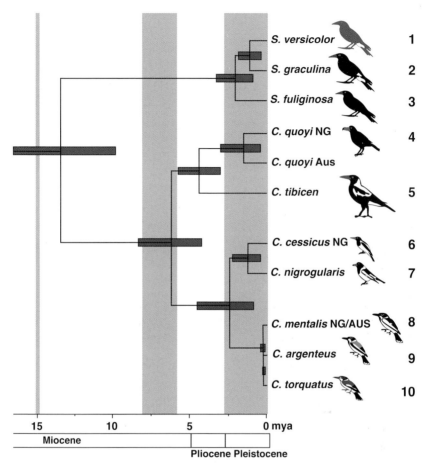

Fig. 1.7. Relatedness of magpies with butcherbirds and currawongs in Australia and New Guinea as presented and conceived by Kearns *et al.* and published in 2013.[24] This chronogram shows the possible diversification nodes and the relatedness of species: 1–3 are currawongs (grey, pied and black) and 6–10 are butcherbirds with species in Australia and New Guinea; 4, *Melloria quoyi* black butcherbird, sub-tropical/tropical; 5, *Gymnorhina tibicen* Australian magpie, Australia-wide distribution; 6, hooded butcherbird, New Guinea; 7, pied butcherbird, Australia-wide distribution; 8, black-backed butcherbird, Cape York only; 9, silver-backed butcherbird (top-end and Kimberleys); 10, grey butcherbird, most of Australia. For our purposes, one of their main findings is that the black butcherbird of New Guinea and northern Australia is clearly identified as the closest relative of the magpie. Note, the magpie (No. 5. above) is not drawn to scale but enlarged for easier identification. According to the timescale constructed by Kearns *et al.*, most speciations would have occurred mainly in the Pleistocene. (Figure adapted and simplified from Kearns *et al.* Latin species names used in the original figure have been retained.)[24]

Other studies further cement the interrelatedness of New Guinea species, not just a subspecies of the magpie, but a family of peltops with Australian woodswallows, butcherbirds and magpies and, to the west and to Africa drawing in the vangas and African bush-shrikes.[25] For example, in 2013, a team of researchers who undertook an analysis of the speciational history of the Australo-Papuan butcherbirds and allies (including magpies), found that the New Guinean black

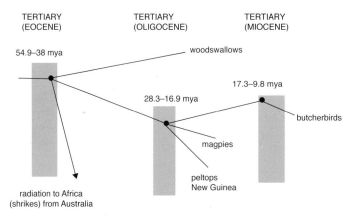

Fig. 1.8. Summary of possible time scales for cracticid/artamid diversifications and radiations to Africa and New Guinea (condensed from latest proposals). Note that all these time frames overlap suggesting the possibility that diversifications might have happened within shorter time-spans than the graph above might imply. Currawongs (not shown here) appear to be less closely related to magpies than butcherbirds but seem to have evolved earlier than butcherbirds, according to Kearns and colleagues.[22]

butcherbird, *Melloria quoyi* (*C. quoyi* in Fig. 1.7), is closely related to the Australian magpie, *Gymnorhina tibicen* (*C. tibicen* in Fig. 1.7).[24]

Interestingly, as Jønsson *et al.*[26] reported, the highest current diversity of 'Corvids' is on the island of New Guinea, where a maximum of 93 species co-occur within a relatively small area. The researchers concluded, based on biogeographical analyses, that the group originated on islands such as New Guinea, probably in the Oligocene/Eocene era and radiated from there.[26] We also know that the latest ice ages did not just affect sea levels but ambient temperature (up to 7 degrees cooler in Sunda and northern Sahul than it is now).

Still, to the person not very interested in the taxonomy and the technical parts of molecular analyses, this tends to be confusing. I have prepared a kind of mud map to summarise the links that this chapter discussed (Fig. 1.8) to show and integrate the African connection to the west and the Papua New Guinea extension of species to the north.

Alternatively, it is quite enjoyable just to look at the birds to which the magpie is supposed to be related (Fig. 1.9). It is difficult to resist the impression that the species below do look rather similar, although we know very well that the first attempts by European explorers and taxonomists to name Australian birds on occasion could be rather disastrous because appearance was often taken as relatedness. For instance, the Australian magpie was named because it looks a little like the Eurasian magpie, *Pica pica*. *Pica pica* is black and white and clever, to be sure, but it has a different beak, different body shape and is no relation of the Australian magpie; neither is the black-billed magpie (*Pica hudsonia*), also known as the American magpie. Today,

Fig. 1.9. Related species: A, the black butcherbird (*Melloria quoyi*) that occurs both in New Guinea and the Top End, distribution as marked in Fig. 1.6); B, Australian magpie; and C, the New Guinea lowland peltops (*Peltops blainvillii*). According to the analyses by Kearns *et al.*[22] the black butcherbird is the candidate most likely to be the closest relative of the magpie.

DNA tests and many others are quite sophisticated and determine such relatedness much more accurately than was ever possible before.

The Australian magpie is larger than any butcherbird, peltops or woodswallow. There is no doubt that the magpie is in a weight class all its own and that it can exert more presence and dominance over its environment than any of these other species can. Subjectively, I cannot help but think that the Australian magpie is the most elegant, even regal-looking (Fig. 1.10), of them all and none of the new

Fig. 1.10. Adult male magpie. The imposing, elegant magpie has a special role in the Australian bush.

findings alter the fact that the Australian magpie towers above the rest of its broader avian family in visibility, as well as contact and friendship with humans, having achieved a rare status for any bird – that of an Australian icon.

Endnotes

1. Hancock 2013
2. Rule 2013
3. Ceballos *et al.* 2015
4. Cracraft 2001
5. Edwards and Boles 2002
6. Cracraft *et al.* 2004
7. Cooper and Penny 1997
8. Martin *et al.* 2014
9. Barker *et al.* 2002
10. Barker *et al.* 2004
11. Christidis and Norman 2010
12. Manegold 2008
13. Harshman 2006
14. Norman *et al.* 2009
15. Fuchs *et al.* 2012
16. Müller *et al.* 2000
17. Schwarz *et al.* 2006
18. Mohr *et al.* 2002
19. Duncan 2002
20. Frey *et al.* 2003
21. Bénard *et al.* 2010
22. Nguyen *et al.* 2013
23. Toon *et al.* 2007
24. Kearns *et al.* 2013
25. Moyle *et al.* 2006
26. Jønsson *et al.* 2011

2
Which is the 'real' magpie?

Australian magpies are popular and have won national approval across Australia, having made it also to Emblem status in Western Australia. Magpies have clubs and organisations named in their honour. One can buy trinkets with magpie motifs, choose works of art (excellent etchings and paintings of magpies) and a good choice of sculptures. Apart from personal contact that many enjoy with wild birds, one of the reasons for this multi-level acknowledgement is that magpies are among a handful of Australian avian species that occur almost right across the nation (deserts excepted)[1] and, because of their size and colour, tend to be highly visible.

When European settlers and scientists first arrived in Australia, they attempted to fit the species they found into known European classifications. Even the vernacular names of Australian birds generally reflected European origins. The magpie is no exception, as already mentioned. A good deal of mythology and guesswork often caused confusion. A charming example comes from possibly the earliest European painting of a magpie (Fig. 2.1), produced by Thomas Watling, who belonged to the group of painters called, in the singular, 'Port Jackson Painter' and these are dated 1788–1792. The painting, now held at the Natural History Museum in London, names the bird as a 'Piping Roller' and its inscription reads: 'This bird has a soft note not unlike the sound of a well-tuned flute. It is a Bird of Prey.' We know differently today.

The IOC World Bird List[2] and the BirdLife Australia Working List of Australian Birds[3] are regularly updated and are considered authoritative reference lists for bird taxonomy. These lists currently give the scientific name for the

Fig. 2.1. Possibly the earliest painting of the Australian magpie, this watercolour is estimated to have been produced between 1788–1792 by the 'Port Jackson Painter'. The artist called the bird the 'Piping Roller' and described it as a 'Bird of Prey'. Source: Natural History Museum, London.

Australian Magpie as *Gymnorhina tibicen*, a name which could mean 'bare-nosed flautist' or 'joyful flautist,' depending on whether one considers the Greek or

Hebrew origins of the name. This is an apt name given the magpie's special status as an extraordinary songbird in terms of octaval breadth and depth of repertoire, a quality that distinguishes the magpie from all other cracticids (butcherbirds).

However, magpies have long been considered an interesting group in taxonomic and evolutionary discussions. One issue is the question of relatedness of magpies to other birds. As early as 1914, Leach suggested that magpies should not have their own genus but be included in the genus *Cracticus*.[4] Schodde and Mason classified magpies in the genus *Gymnorhina* and butcherbirds in the genus *Cracticus*, both of which they included in the family Artamidae.[5] Since their work, however, genetic studies have shown that the magpie is actually 'nested' within butcherbirds; that is, the Black Butcherbird (*M. quoyi*) is more closely related to the magpie than it is to other butcherbirds.[6] That opens up several options for generic classification of magpies and butcherbirds. While some publications have opted to refer to the Australian Magpie as *Cracticus tibicen* (e.g. Christidis and Boles[7]), and some lists treat it as *Gymnorhina tibicen*,[2,3] magpie taxonomy continues to be a dynamic area of study in which researchers have not yet reached full agreement.

Given the ongoing nature of the debate, this book follows the current IOC World Bird List and BirdLife Australia Working List of Australian Birds in using *Gymnorhina tibicen* and the family Artamidae when referring to the Australian Magpie. This approach is also in agreement with Cake *et al*.[30]

Generally, one can speak of two obvious plumage differences in magpies: one is called the white-backed magpies and the other the black-backed magpies (Fig. 2.2).

Fig. 2.2. White and black-backed magpie comparison. Left: White-backed magpie (WBM). Right: Black-backed magpie (BBM). Both images depict males. Of all magpies, the white-backed magpie of Western Australia is the most sexually different (dimorphic) – while males of WBM sport a completely white back, females have a visibly different greyish/motley back. In the BBM, sex differences are very slight – the female may have a small strip of grey feathers at the nape of the neck (between the black and white feathers).

The white-backed magpies are largely in the southern and western parts of Australia and the black-backed magpies across the rest of the continent.

Magpies: one or several subspecies?

There has been ongoing debate about the status of the magpie as a single species or several subspecies for as long as the task of classifications of Australian birds has been actively undertaken. Magpies, first classified by Latham in 1802,[8] started out as a monotypic species that was slowly split into subspecies as knowledge grew over the 19th and early 20th centuries (Table 2.1 below). Subspecies status is decided on such features as subtle differences in plumage colour, bill length and other pertinent anatomical features, such as wing length: Latham (1802)[8] for *Gymnorhina tibicen tibicen*, Gould (1837)[9] for *Gymnorhina t. hypoleuca*, A.J. Campbell (1895)[10] for *Gymnorhina t. dorsalis*, Milligan (1903)[11] for *Gymnorhina t. longirostris*, Mathews (1912)[12] for *Gymnorhina t. terraereginae*, H.L. White (1922)[13] for *Gymnorhina t. eylandtensis* and Schodde and Mason (1999)[5] both for *Gymnorhina t. dorsalis and Gymnorhina t. tyrannica*.

There is actually another subspecies, not often mentioned – the New Guinean subspecies of the magpie, *Gymnorhina tibicen papuana*. In 1986, Black[14] stated that the New Guinean magpie *Gymnorhina tibicen papuana* is distinctive in several ways. He noted in particular the exceptionally large bill, which is both longer and deeper than in other forms. In dorsal plumage it is similar to the western magpie *G. t. dorsalis* but in other characters it resembles the long-billed form of the black-backed magpie *G. t. longirostris* of north-western Australia.

It is sometimes fun to follow the source material back to its original field work, in this case undertaken first in the 1920s. At that time naturalists, ornithologists and anthropologists from Europe went out to places like New Guinea and spent considerable time under difficult conditions in the forest to identify and classify species and then, unfortunately, had to kill the specimens to take them back to European museums. Bangs and Peters described the subspecies first originating from Princess Marianne Strait in south-west New Guinea.[15] Other Dutch explorers worked in New Guinea in the 1960s. Fig. 2.3 shows two specimens that were collected and are now held in a Dutch Museum. The studies by Mees in the lowlands of southern New Guinea (Merauke and Koembe) first described a white-backed adult male magpie and then an immature black-backed one (Fig. 2.3).[16] The description of his finds are given verbatim in the caption.

This New Guinea magpie, interestingly, shares morphological traits with subspecies from north-western and south-western Australia, *G. t. longirostris* and *G. t. dorsalis*, respectively, as said before, not with magpies in the northern territory or in Queensland's north. The most recent work, using the latest techniques,

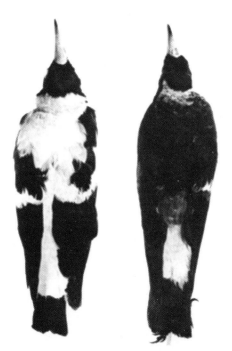

Fig. 2.3. New Guinean magpie. Right: 62 mm, weight 328 g. Iris very light brown, bill pale slate-blue with a few dark stripes near base on maxilla and a dark tip, legs black. No moult, plumage moderately worn, broad black saddle on the middle of the back, posteriorly bordered by grey; rump and base of tail white; nape white'.[16] Left: first adult male known of this race, which unexpectedly was found to have a white back (cf. Mees).[17]

mitochondrial DNA and Z-linked DNA, has confirmed this close relationship with the two western Australian subspecies,[18] suggesting movement of magpies between New Guinea and mainland Australia (possibly in both directions) at the very latest during the last glacial period that ended ~16 500 years ago when sea levels began to rise again but it could have happened in an earlier period. It also suggests that the New Guinean magpies did not settle in the Cape York Peninsula,[18] but eventually made their way to the west and it is from that population some must have made the trans-ocean voyage to Africa, as was described in the previous chapter.

The distribution of magpies according to subspecies designation shows also some considerable overlaps as marked (Fig. 2.4). And, of course, as subspecies they can also breed with each other and produce viable offspring.

As can be noted, plumage patterns and sex differences expressed in plumage differences on either the back or the nape of a female are important identifiers when studying them in the field.

In recent years, the number of reports that I have received about sightings of magpies diverging from the expected plumage patterns and colours has increased. Magpies have been sighted that are almost completely white (without albinism),

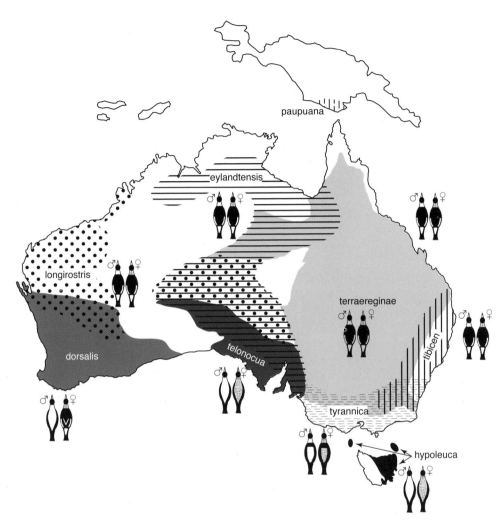

Fig. 2.4. Distribution of magpies in Australia. Latin names of subspecies are listed in each marked region of their occurrence; striped areas indicate that two or more adjacent subspecies can be found. It is seen clearly here that white-backed magpies occur in the south of the continent and in Tasmania, while black-backed magpies occupy the rest of Australia to the very north of the continent, including also the ninth subspecies of *G. t. papuana*. Unfortunately, the specimens that have been tagged, as Fig. 2.3 shows, are inconclusive (either white-backed or black-backed). Adapted from and based on Schodde and Mason.[5]

others whose adult plumage patterns have exaggerated scalloping and yet others that have near normal plumage patterns but have brown rather than black plumage (Fig. 2.5). Changes in pigment levels can occur (more about melanin and other pigments in Chapter 3). Whether these are aberrant changes triggered by something in the environment (be these toxins or diseases) about which we should be concerned or whether some of the changes may be patterns emerging from

Fig. 2.5. Plumage changes. A, magpie has brown scalloped chest feathers (see also Plate 3). B, total loss of black and almost total loss of any colour (the facial colour is a light brown). C, while the plumage is the usual black and white, its distribution and patterning are highly unusual, especially on the head.

geographically overlapping subspecies is not known. It is certainly worth watching and noting how widespread these occurrences are.

The different subspecies vary considerably in size, shape and even plumage patterning depending on the region and the geographical conditions.[19]

The largest magpies live on the east coast from Melbourne to Brisbane. Top End magpies, *G. t. eylandtensis*, have long and slender bills, Tasmanian magpies, *G. t. hypoleuca*, have the shortest and most compact. The differences, so Alex Milligan[20] explained, were explicable in the different soil and climatic conditions and that these conditions led to different adaptations. In addition, one suspects that the long and slender bill of Top End magpies (for both *G. t. eylandtensis* in

Table 2.1. Australian magpie subspecies: morphology and distribution.

Subspecies	Size	Beak length (mm)		Wing span (mm)		Geographical area
		Male	Female	Male	Female	
G. t. eylandtensis	small	56–62	51–57	230–255	225–245	NT
G. t. terraereginae	small–medium	48–58	47–53	245–265	235–255	Qld, inland NSW, Vic. and SA
G. t. tibicen	large	48–55	45–50	260–385	255–270	Brisbane–Vic. border
G. t. tyrannica	very large	52–57	47–53	270–290	260–280	Vic. and into SA
G. t. hypoleuca	small	43–47	38–43	248–258	235–245	Tas.
G. t. telonocua	medium	50–56	45–50	255–265	245–255	coastal SA
G. t. dorsalis	medium	56–60	48–54	258–270	240–255	south-western WA
G. t. longirostris	medium	60–65	55–60	245–260	235–250	north-western WA

Source: Schodde and Mason[5]

the Northern Territory and *G. t. longirostris* in the north-west of Western Australia) is not only related to soil conditions, as Milligan argued, but also to the type of food found above the ground. Northern and north-western magpies in long periods of very dry weather may encounter many scorpions and other poisonous fare (e.g. spiders) and may have added these to their menu. The very short and compact beak of the Tasmanian magpie, on the other hand, may be more suitable for cracking the body armour of hard-shelled beetles and cockroaches. However, this is speculation as we have no data so far to confirm Milligan's conjectures.

Body size differences between subspecies of magpies are not so easily explained by climate alone. If all tropical and subtropical magpies were small and all temperate climate magpies medium and large, one would most likely infer climatic reasons to explain the size differences (called Bergman's rule). While the tropical ones are small and medium in size, *G. t. eylandtensis, terraereginae* and *longirostris*, the size differences between subspecies at the southern tip of the continent do not conform to a climatic model. In Tasmania, *G. t. hypoleuca* magpies are small, while those facing Tasmania from the mainland are particularly large, *G. t. tyrannica*. This means that climatic conditions, although not identical but very similar between southern Victoria and Tasmania, cannot account for the most substantial differences between magpies almost anywhere on the continent, although smaller sizes are typical for islands (such as Tasmania). As already mentioned in the previous chapter, DNA analyses and the possible history of radiation shows that some magpie groups moved west and have more in common with eastern magpies than magpie subspecies that may live in adjacent geographical spaces today. Such movements may have been prompted by climatic events and episodes of extreme aridification of inner Australia. According to reliable dating techniques, the time of the last cold phase of the Quaternary Ice Age happened ~18 000 years ago.[21] While such ice ages apparently had relatively few consequences for Sahul, it did result in climate changes that contributed to a drying out of the continent and to refugia in arid landscapes during the Pleistocene. Toon and colleagues showed that this aridification also played a role in structuring populations of the Australian magpie.[22]

Taxonomists devise current classifications on the basis of several criteria in order to arrive at their conclusions. Over time, these may hold or be overthrown. Some of the difficulties are well demonstrated in magpies.

How important are the differences in the magpie subspecies?

In these taxonomical debates, one may distinguish between primary, secondary and trivial differences but sometimes it may be difficult to tease out the causes of the differences and determine their relative importance. What, for instance, could be the reason for the variations of the markings of the plumage, particularly on the back? Richard Schodde and Ian Mason argue that all magpies are probably derived from two forms: the white-backed form as Bassian (southern/temperate) in origin and the black-backed form originally Torresian (Top End/tropical and subtropical)[5] but there are now more recent findings by Toon *et al.* showing that all populations within the eastern region are clustered together, while all populations within the western region are 'as divergent from each other as they were divergent from all populations within the eastern region'.[22]

Size and bill length may also not be explicable solely by geographical variables. Other differences may include colour or pattern differences as these are found in tail and wing patterns. For instance, in magpies tail bands may vary in size, be this by subspecies (e.g. in *G. t. eylandtensis*, the black terminal tail band is narrow while in *G. t. tyrannica* it is very broad), by age (broader in young birds) and sex (females have broader tail bands). Finally, the black versus white backs of magpies, although the most dramatic and obvious of all differences (at least to the casual observer), may be a minor difference in terms of species designation (but not necessarily 'minor' for individual magpies).

Although subspecies status may be delineated on subtle differences in plumage colour, bill length and body size[5] there may, in fact, also be behavioural differences that arise from different environmental demands. Behavioural criteria are usually not included in taxonomic descriptions as they could be adaptations to specific environmental criteria (although some behaviours are considered important such as nest building and nest type). In this regard, the study of magpie behaviour could be cause for further questions of the magpie's species status. Foraging strategies and nest-building type are the same for magpies across the continent but social organisation is not. There are other behavioural indices, such as song, to be discussed in Chapters 9 and 10, that magpies have similar skills but use them in different ways.

Magpies, like the majority of songbirds, are normally found in either groups or pairs. Being sedentary and altricial (confined to the nest for a period after hatching) means that magpies, if bred in good permanent sites, are raised by at least two adult birds and grow up in a stable environment. Magpies that live as solitary individuals have usually been orphaned as nestlings or fledglings, injured (see Chapter 9) or evicted from a territory (see below). Any of these circumstances can cause a fatality and solitariness can be a contributing factor to an individual bird's demise.

Sizes of territorial groups vary with geography and with habitat variability and quality.[23] Group sizes of non-breeding individuals vary according to the ages of its members, the quality of feeding grounds, particular seasonal rhythms as well as to geographic location. Therefore, group size and territorial size are not necessarily correlated.

Group size has been extensively studied in Western Australia,[24] in Adelaide, South Australia,[25] and in Canberra,[26] and more recent studies have added new sites in Western Australia, the Nullarbor Plain, Melbourne, Tasmania and Queensland, including Brisbane and towns north of Brisbane such as Nambour, Rockhampton and Townsville, as well as one location (Brunette Downs) in the Northern Territory.[23]

As indicated above, numbers of magpies per group vary with seasons. Group size increases as a result of breeding and then declines before the next breeding season. The degree of the decline depends on several factors including attrition rate of the young and group behaviour towards juveniles. In Queensland, it appears that juveniles regularly disperse towards the end of the previous breeding season[23] while in Canberra[27] and Western Australia[24] the young are often permitted to remain within the natal group and they do so often for as many as three years.

In one of my own research sites in the New England area, the young were often permitted to stay while the previous occupants (a pair) had ousted their young from their territory before the onset of the next breeding season. It seems, that there is no hard and fast rule about this social behaviour. However, the behaviour may be determined by available food resources at the time and these can fluctuate between seasons and for the same season from one year to the next.

The various studies have revealed some substantial and significant differences in magpie group sizes in different parts of Australia. Not counting juveniles in any of the samples, in Queensland, the Northern Territory, the sparse Nullarbor Plain and in some parts of South Australia permanent groups can range typically between two to five individuals. In the regions around Canberra, Melbourne and the Great Divide between Armidale and Coffs Harbour, magpie groups typically range from two to ten individuals. By contrast, in Tasmania (Launceston) and in Western Australia (Coolup, Perth) groups are very large, ranging from three to as many as 15 or, in Coolup, 26 individuals.

It would be tempting to hypothesise that the smallest magpies gather in the largest groups. Tasmanian magpies, are the smallest magpies of all in Australia and they do, indeed, consistently form the largest groups.[23] However, this model cannot be upheld for all subspecies across Australia. The next smallest magpies (in the Northern Territory and north-west Queensland) have particularly small groups (usually only pairs) and the largest magpies, such as *G. t. tyrannica* (Victoria) and *G. t. tibicen* (east coast), have medium-size groups.

The conclusion that Hughes and Mather drew in their study published in 1991 is that group size varies with habitat quality.[23] Perhaps 'habitat quality' requires closer definition. It is not just the fertility of the soil and the richness of insects, larvae etc. that this soil may provide but 'quality' of habitat also depends on accessibility of food and water. For instance, long grass may require more territory than short grass as a study on magpie territories found.[27] Short grass, incidentally, is best produced by native grazers such as kangaroos or wombats, not by cattle or sheep because the latter tend to take out the roots of the plants and compact the soil due to weight and the structure of hoofs. The relationship between nutritional values, foraging and territorial size will be explored later. Suffice it to say, that group size relates to another aspect of social organisation, namely the likelihood that a larger group will breed cooperatively rather than as a pair on its own. Thus group size may suggest traits that are adaptive and have been practised and especially well maintained in certain populations (for instance in *longirostris* and *dorsalis* – see Fig. 2.4). Populations in Western Australia may have been segregated from other populations for a long time. Equally, dispersal patterns may be cited because they too are regionally different.[28]

Today, although the question is still not entirely settled, taxonomists have reintegrated all the magpie variations into one species by reassigning those of former separate species status to subspecies status. However, the number of separate subspecies has grown from six to eight to include two new subspecies *Gymnorhina tibicen tyrannica* and *G. t. telonocua*.[5] These subspecies, all nine of them when the Papuan magpie is included, are not sharply divided in most regions, as already said, because mixed forms may occur in adjoining regions from one subspecies to another (see Fig. 2.4).

In summary, if there are variations in breeding strategies, group size or plumage, for instance, the studies have shown that these are a matter of geography, possibly effects of refugia during ice ages and even climate. Yet after describing all the important detailed studies on magpie variations across the continent, it is all right to just speak of 'the magpie' wherever it may be.

Endnotes

[1] Barrett et al. 2003
[2] Gill and Donsker 2018
[3] http://www.birdlife.org.au/conservation/science/taxonomy, version: 2.1
[4] Leach 1914a
[5] Schodde and Mason 1999
[6] Kearns et al. 2013
[7] Christidis and Boles 2008
[8] Latham 1802 in Schodde et al. 2010
[9] Gould 1837
[10] Campbell 1895
[11] Milligan 1903
[12] Mathews 1912
[13] White 1922 cited in Cambell 1929
[14] Black 1986
[15] Bangs and Peters 1926
[16] Mees 1982
[17] Mees 1964
[18] Toon et al. 2017
[19] Hughes 1980
[20] Milligan 1903
[21] Linacre 1999
[22] Toon et al. 2007
[23] Hughes and Mather 1991
[24] Robinson 1956
[25] Shurcliffe and Shurcliffe 1974
[26] Carrick 1963, 1972, 1984
[27] Carrick 1972
[28] Hughes et al. 1983
[29] Baker et al. 2001
[30] Cake et al. 2018

3
Anatomy

One of the obvious and self-evident descriptors of a bird is that it has feathers and can fly, or it once had the capacity to fly. When we say 'bird' we immediately have a fairly clear idea of its anatomical features – two legs, two wings, feathers and a beak. But the evolution of birds has produced an interesting paradox resulting from two different sets of forces.[1]

One is defined by the evolutionary adaptations for flight. The ability to fly imposes several very definitive anatomical design restrictions and these are reflected in uniformity of structure. There are many such restrictions – bones must be strong but not heavy, the body must be streamlined and light (helped by several air sacs within the body). Forelimbs had to be converted to wings and had to serve almost no other function than flight. The head had to be light and the brain weight kept down. Legs needed to be able to fold away during most types of flight and the integument required a surface that could take pressure and turbulences without toppling the bird. Adjustments of the senses (especially of vision and of the sense of balance) and of the respiratory and circulatory system are all important adaptations for flight. Flight uses far more energy than any other form of locomotion and this too has had an impact on the overall anatomical design of birds.[2]

The other force affecting the structure of birds is ecological. Being able to fly has resulted in a degree of adaptive radiation that other orders usually do not share. Birds have colonised just about any habitat on Earth, including islands. This means that they have developed a far greater number of variations in terms of feeding and locomotion, colouration and even metabolism related to their respective niches than other orders – there are over 10 000 verified species of birds alone[3] as against less than half that number of mammals. Hence, the paradox is that achieving flight imposes severe anatomical restrictions while, at the same time, substantially expanding the range of possible habitats and thereby the degree of adaptive radiation (adapting even anatomically to these habitats). Restrictions in this case produce a higher degree of variation – this explains the tremendous number of bird species.

Although some basic design features are present in all birds, there are many important, though sometimes subtle, differences of anatomy between species. This chapter will not attempt to outline all features that are part of a bird's external and internal anatomy but only briefly refer to observations of the skeletal system and the integument (its external surfaces) when these may be able to say something specific about the ecological niche occupied by, and may partly explain the behaviour of, the Australian magpie.

Integument

The external surfaces of a bird include the skin, feathers, bill, claws and any protrusions. Magpies do not have any of the latter – no comb (as in chickens), ricti (folds at the corner of the mouth), wattles (as in wattlebirds), or snoods (fleshy parts hanging from the nostril, as in turkeys). The surface of magpies and all artamids is smooth and covered by feathers, except for the beak, the legs and feet.

The skin of birds is vascular: it is much more fragile and a good deal thinner than in mammals and it has few connections even to connective tissue and more to bones than in mammals. Muscles are interconnected by tendons, which connect to feathers. The skin of magpies has no true cutaneous glands, except for glands in the external ear, around the vent and a specific gland situated dorsally near the tip of the tail, called the preen or uropygial gland. This gland produces a lipoid sebaceous fluid that the bird collects with its beak during preening. It then draws individual feathers through its beak and covers them with this fluid thus waterproofing them, at the same time keeping its feathers, scales and beak supple. In heavy rainfall it is apparent that magpies do not seem able to waterproof the crown of the head well, because these feathers tend to get soaked.

Magpie skin does not have sweat glands; indeed, no other birds have sweat glands either (although water loss through the skin has been observed in some species). There is some cutaneous heat loss brought about by changes in skin blood flow and that can be partly fostered by the lifting of wings. Most heat control is usually not achieved via the skin but by feather positioning and, largely, by rapid respiration in case of high heat load.

Ears and eyelids are also part of the integument. Eyelids are closed from the bottom lid upwards and this occurs usually only in sleep and when the individual

Fig. 3.1. Magpies have an 'all purpose' beak: solid, sturdy, nicely V-shaped and useable for fighting, feeding, breaking open hard soil and crushing even harder-shelled fare, such as beetles. Nestlings (left) have a much shorter beak, lengthening to adult size (right) by about three months of age. Note that the beak of a nestling is much darker than that of an adult, usually slate, and the eyes of juveniles are dark brown while those of adults are light brown with a reddish tint. Beak, feathers and eyes contain melanin pigments.

bird is in pain. Birds also have a translucent third eyelid, called the nictitating membrane, that traverses the eye from the lower inner part across the eye. The membrane, in keeping the surface of the eye clean and moist, may move rapidly many times per minute and thus takes on the function of 'blinking' that we see in mammals.

Beak

All artamids have hard and relatively thick beaks. There are about 16 main designs of beak in birds, with a multitude of modifications and each form is congruent with the feeding niche that a species has evolved to occupy (see next chapter). The beaks of magpies fall into the category of 'all purpose' beaks, as do those of ravens. However, unlike the beaks of ravens and of currawongs, magpie bills are more slender and more pointed at the tip (Fig. 3.1). For a true ground feeder with extractive foraging habits the beak is ideally suited for piercing the ground, also capable of driving a wedge into relatively hard soil. The beak of a nestling is relatively soft, too soft to pick up, crush or pierce any food and the wide gape is also flexible and too soft to grab and hold any food, as nestlings of some other species can. Hence the nestlings need to be fed to the throat rather than to the beak. As an 'all purpose' beak, it has of course more functions than this and reflects the fact that magpies use a variety of food types as shall be discussed in the next chapter.

Typical for all artamids is the greyish colour of the beak with a darker tone at the very tip of the beak. All artamids use the beak to catch and kill a prey item. In magpies and butcherbirds beaks are also used effectively in fighting. They can injure an adversary or, in rare and extreme cases, even kill an intruder or unacceptable fellow magpie.

Feathers and feather colour

Feathers are enormously versatile. They protect birds from heat and cold, light and rain, as well as enable them to fly. Depending on the position of the feathers (sleeked down, fluffed or raised) body temperature can be increased or decreased. Feathers are a great insulator and the preening gland secretions maintain their suppleness.

Magpie plumage, as the majority of corvoid plumage anywhere in the world, has the singular distinction of being largely confined to black and white and shades of grey. Birds have different pigments in their skin and feathers.

The majority of songbirds, especially those of the superfamily group of Corvoidea, have the distinction of being the main wearers of black and this is achieved by a pigment called melanin. Melanin can occur in the skin, in the eyes and at three different layers of a feather. When most melanin is concentrated at the outer layer (cortex) of feather, the feathers appear intensely black, the further down in a feather it is (be this in the middle layer, called the cloudy layer, or at the inner core, the medulla) the lighter the appearance will be, looking grey so that the colour differences on the nape of the neck or the back may not always be an indication of the density of melanin but about the position of melanin within a feather. Melanin can also result in brown hues. Ravens and many crows are intensely black, so are the black butcherbirds of Papua New Guinea and the pied currawong. Certainly, we perceive the black feathers of a male adult magpie, especially on the wings and chest, as intensely black.

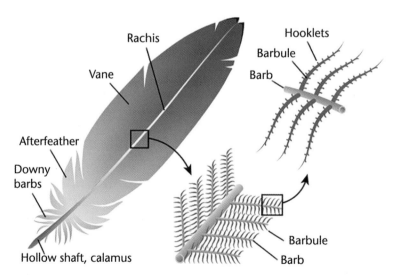

Fig. 3.2. The structure of a feather. Melanin is deposited in the barbs and barbules of the feather at the time of growth of the feather. Once that has happened, no new supply of melanin reaches the feather for the length of the feather's life. (Drawings from Ask a Biologist,[4] CC BY-SA 3.0)

Fig. 3.3. Adult male at breeding time. Note the gleaming black feathers, a clear indication of health. The author shared a property with this bird and his family for 20 years. Although the photo was taken a mere metre away from the bird, this male is relaxed, in good health,[5] confident and enjoying the winter sunshine. At no time did this male ever attempt to restrict my movements, let alone swoop.

The reason for making any reference to melanin here at all is that in recent years a good deal of research has been undertaken to show that the presence of melanin and its density can tell us a great many things about an individual bird and its environmental context. One is the overall health[5] of the individual (Figs. 3.2 and 3.3). One study showed that eumelanic individuals (those with brown or black hair or fur) are more immunocompetent, resistant to parasites and developmentally stable.[5] Interestingly, high density of melanin reduces preening time so that energy can be directed elsewhere and often lead to more effective vigilance behaviour and even confidence.[6] It was also found, for instance, that behavioural traits such as exploratory behaviour are linked to melanin levels and, as shall be explored later, even to brain asymmetry.[7]

These colours raise several complex points about the way in which they may interact with the environment, i.e. whether one colour or the other benefits the individual in certain contexts. The presence of melanin may make the feathers more resistant to wear and tear than white or other colours do. White feathers, indicating the absence of melanin, may be useful in habitats with extensive exposure to sun although black ravens seem to be no worse for wear even in hot regions. Very rarely, magpies have been found with almost completely white plumage.

Feathers may also be used for display to attract a mate. There is no evidence to suggest that plumage colour in magpies has a role to play in mate choice. Indeed, in some cases, especially in territorial species as the magpie, looks may be of far less importance than a good property/territory.[8] Of course, a healthy bird that is not suffering from parasite overload will have a gleaming coat and this may be of some importance. In other species it has certainly been shown that plumage condition is a factor in female mate choice[9] but this is less likely to apply to magpies (see Chapter 7).

Feathers often have the additional function of providing camouflage. Black or shades of grey may work very well as camouflage. Ravens (black) or woodswallows (shades of grey) may be able to merge into the background of their habitat. Australian magpies, by contrast, are striking in their black and white contrast plumage and conspicuous. They can be spotted easily in any landscape they have colonised.

Such extravagance may have several adaptations, be this in courtship displays[10] or for camouflage. Some camouflage, obviously, is not just for the purpose of protection *from* predators but offers a disguise *for* a predator. It suggests, as indeed we know, that magpies are not ambush hunters. Magpies are almost exclusively ground feeders, like quail and chickens, but, unlike most other ground feeders that are predominantly prey species, magpies can apparently afford to be seen. Their greatest vulnerability, as for most songbirds, occurs during the breeding season. It is not the adults, but the nestlings and juveniles, that are at risk from predators. Once they have reached adulthood, magpies have few enemies and it is the adult plumage that dares to be particularly conspicuous.

Visibility, among any species, may well be a sign of dominance and perhaps ought to suggest a lack of significant predator species in their habitat. Interestingly, in other species female birds are often well camouflaged when they nest in the open. Magpie females, solely responsible for incubating the eggs, may be considered to have some camouflage – even among the so-called white-backed magpies, females actually have a mottled back.

Wings and tail feathers

Magpie flight feathers (called remiges) follow the general pattern of wing construction and have two main divisions. The main flight feathers are attached to the manus (carpometacarpal bones and phalanges). The secondary feathers are attached to the ulna. The number of primary feathers of flighted birds can

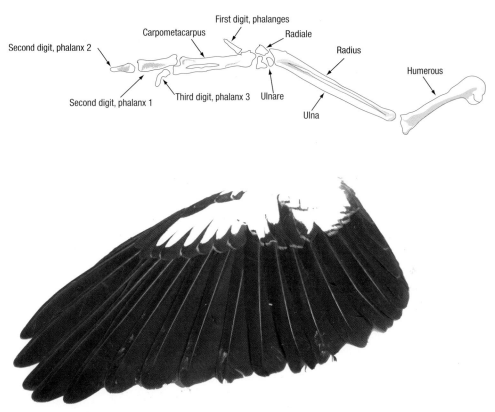

Fig. 3.4. Left wing of an adult magpie, showing both skeletal (top) and feathered (bottom) features. Not all feathers are shown.

vary between 9–12 primaries. Most songbirds, including the magpie, have 10 primary feathers. The number of secondary feathers varies a great deal between species depending on length of forearm and flight needs (from six to as many as 40). Magpies have 11 secondary feathers and these follow on from the primaries without a gap. This feature is not shared by all avian species and it promotes quiet, strong flight.

Tail feathers, called the rectrices, are effectively used in flight for stability and steering and also act as brakes when landing. In magpies, tail feathers are used for intimidation displays when vanquishing an enemy or play mate. There are 12 rectrices in a magpie tail – largely white but rimmed at the tip of the tail feathers in black. They can be splayed in a perfect semi-circle, looking very much like a handmade fan. Damaged tail feathers can only be replaced during moult (once a year) but if any tail feathers are pulled out completely, they may grow back within six weeks. Flight is affected in such circumstances but usually the bird can compensate for the loss of tail support by using the wings as the sole control.

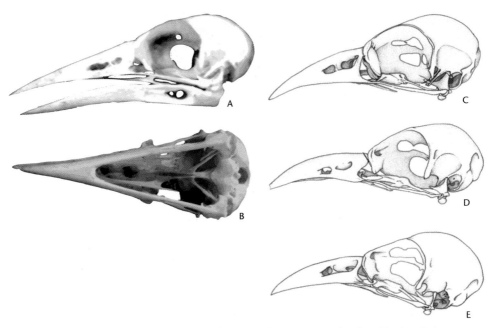

Fig. 3.5. Comparison of the skeletal features of a magpie head (photographs of A, side view: B, from below), with those of related birds (C, white-breasted woodswallow, *Artamus leucorynchus*; D, mountain peltops, *Peltops montanus*; E, Chabert vanga, *Leptopterus chabert*). Note the magpie's beak features, large eye socket and skull to hold the large brain and bones of the palate. Though the images are not to scale (woodswallows and the New Guinean peltops are much smaller than magpies), there are some clear similarities. The vanga (E) shown here occurs in Madagascar. (Source: C–E, Manegold).[11]

Head

In the skeleton of birds, it is the head that varies the most among species. Not altogether surprisingly, it is the head, among other features, that gives artamids their very special family status. Particularly in the palate and in the orbital regions (above the eyes), artamids show a set of traits that are unique to this family. The narrowed palate across all artamids is phylogenetically significant because of the very different feeding habits of the members of this family.[11] Woodswallows may feed in trees and on the wing, butcherbirds catch large insects on the ground, and currawongs are largely fruit-eaters. Only the magpie uses its beak to pierce the ground and extract food from below the surface. Usually different feeding habits lead to anatomical adaptations, but here, the narrow palate has been maintained.

The morphology of the magpie's skull has been of interest in studies on relatedness of various species, and now is perhaps the best moment to come back to Manegold's study in speciation of the Australian artamids.[12] As was discussed in Chapter 1, Manegold in particular pointed out that there are links between the New Guinean peltops, Australian artamids, including the magpie, and the African vangas.

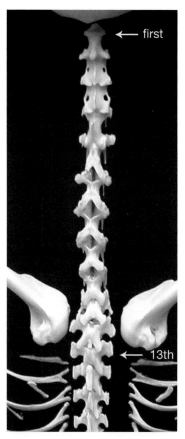

Fig. 3.6. Vertebrae of an adult magpie. The vertebrae are counted from the head (top of image) to the beginning of the chest (first and last neck vertebrae are marked by white arrows).

Fig. 3.5A,B shows the actual skull of a magpie that I sourced from the University of New England as a record of the anatomy of avian species. The skull is part of its Natural History Museum's collection. This is simply a photograph while the skulls of other species presented in Manegold's research paper are careful and detailed drawings (Fig. 3.5C–E). Nevertheless, despite the obvious inadequacies of a photograph, at a superficial level at least, the similarity between the skulls is impressive.

Neck

Birds have more cervical vertebrae than mammals, and this generally gives them a greater degree of mobility in the neck region than in mammals. This additional mobility is needed because eye movement tends to be very limited and so the mobile neck compensates for this, allowing visual scanning of the environment. A mobile neck is also specifically necessary to reach the preening gland.

Vertebrae are subdivided into three main sections. Those in the area from the head to the first complete set of ribs are called cervical vertebrae. These are followed by the thoracic vertebrae (with the rib cage) and then the lumbar vertebrae, which are towards the rump (not shown in Fig. 3.6).

In mammals, there are usually only seven cervical vertebrae, while in birds they range from 11 to as many as 24 (in swans, for instance). The magpie has 13 cervical vertebrae (see Fig. 3.6). The length of the neck depends on the number of cervical vertebrae. Some long-necked waterbirds and shorebirds with exceptionally long necks have correspondingly high numbers of cervical vertebrae. Usually, neck length and length of feet are matched.[1] The magpie is well proportioned, with the length of the legs and neck being about even.

Sternum and ribs

The sternum, or breast case, is much more highly developed in birds than in mammals, and is one of the important taxonomical elements in the classification of birds. From the sternum protrudes the keel, a bony plate almost at a right angle, like the keel of a boat. This is very well developed in the magpie. Some birds, such as the domestic chicken, have notches or windows in the sternum but these are absent in magpies. The strong keel and the absence of notches and sternum windows suggest a strong flyer – it can support large pectoral muscles used in flight.

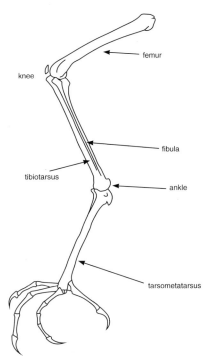

Fig. 3.7. Leg of the magpie. Note that the knee, located between the femur and the fibula, is not visible in a live magpie. However, the ankle (intertarsal) joint is visible and at times mistaken for the knee.

The number of ribs may vary, according to species, between three and nine pairs of 'true' ribs (full-length ribs). The magpie has six pairs of full-length ribs and these protect internal organs.

Legs

The legs consist of several sections. The bone articulating with the hip is the thigh (femur), followed by the knee joint. The lower leg (fibula and tibiotarsus) ends with the ankle (intertarsal joint). Below this is a relatively large section, at least in magpies, called the tarsometatarsus, followed by the foot with its four toes (digits). Usually, what we see of a magpie's leg is only from the ankle joint downwards (the tarsometatarsus) because the thigh is hidden in the body and feathers reach down to the knee joint. When the birds sits down it thus folds the legs into almost equal halves, shifting the centre of gravity back for balance against a forward head and neck.

Paeleoanthropologists and biologists look at leg bones of extinct animals with great interest because the legs, and the way of their articulation with the hip, can tell them something about the locomotion of a species. A long thighbone (nearly or of equal length as the lower leg bone), for instance, indicates that the animal in question was a fast runner. A short thighbone, by contrast, generally indicates that the species is predominantly a walker. In magpies, the femur is substantially shorter than the tibiotarsus, giving a clear indication that the bird is not designed for habitual running but for walking. Magpies can run and do so for short distances, sprinting when they see a food item a little distance away on the ground.

It appears that the magpie is the only cracticid species that has developed as a fully bipedal walker, in the manner of shorebirds and waders. By contrast, all other species in this family predominantly hop. Butcherbirds, woodswallows and even currawongs tend to keep both legs in parallel position and hop from one place to the next. Ravens, like magpies, have also developed a gait but ravens need to move the side of the hip forward together with the leg they move so that their gait is a slight shuffle with an alternating body twist forward in line with whatever leg they are moving at the time. It gives their gait an awkward clumsy appearance and keeps their legs firmly apart. In magpies, the legs are placed relatively close to each other and the gait is an easy forward motion without any tilting of the body.

This method of locomotion in magpies is consistent with their feeding behaviour. Ground feeders, including shorebirds and waders, are true bipedal walkers. All other birds that are not exclusively or predominantly ground feeders have obviously had no need to develop the anatomical features necessary to achieve a bipedal walking gait.

It is useful to remember true bipedalism (putting one foot before another and sustaining such a walk for long periods) is a rare attribute in the animal kingdom. Some species, such as lizards and a few mammals, have developed a running

Fig. 3.8. Feet of a magpie. Image taken from below a glass-top table splendidly exposing the feet, as they are rarely seen. It shows how large the feet are and how well padded for a life of walking and roosting. (Photo: The Magpie Whisperer, www.magpieaholic.com)

bipedalism at high speed; most songbirds are hopping bipedalists.. Indeed, only humans have completely adapted to bipedalism since it first fully developed in birds (and possibly some dinosaurs before them).

Feet

There are about seven basic types of feet in birds. One distinguishing characteristic is whether the feet are fully or partially webbed or not; another is the number and arrangement of the digits (toes). All birds have either three or four digits and many species of those with three digits often still have a fourth vestigial one. All songbirds, including the Australian magpie, and all birds of prey have unequally positioned digits referred to as an anisodactyl (unequally toed) foot (Fig. 3.8).

Three toes face forward and the fourth backwards (see Fig. 3.7). In songbirds, the three forward digits are usually placed rather close together while those of raptors are spaced widely apart. All digits are freely mobile (which is not the case in the kookaburra where the third and fourth digit are partly fused at the lower end, called a syndactyl foot), and the digit facing backwards is opposable. This is a hallmark of perching birds. In birds of prey, the same digit is also freely opposable.

It is perhaps a little odd to finish a chapter on one of the foremost songbirds in the world by talking about its feet! There are some ways in which feet can lead back to a general discussion about magpies. One important fact is that in some avian species, including all parrots and all birds of prey, the brain has extra circuits for foot/brain coordination. Magpies do not scratch the ground in search of food as do chicken, lyrebirds or megapodes, they do not catch their food with their feet as all birds of prey do but, in exceptional circumstances, they may hold food with the left foot in order to steady the item. If, as I suspect, such action tends to always occur on one side (most cockatoos are left-footed), this is called 'footedness' and relates back to brain organisation and will be explored further in the next chapter.

Endnotes

[1] King and McLelland 1984
[2] Baumel 1993
[3] Gill and Donsker 2017
[4] https://askabiologist.asu.edu/explore/feather-biology
[5] Jacquin *et al.* 2011
[6] van den Brink *et al.* 2012
[7] Mateos-Gonzalez and Senar 2012
[8] Alatalo *et al.* 1986
[9] Bennet *et al.* 1997
[10] Galván 2008
[11] Schodde and Mason 1999; see their discussion p. 532
[12] Manegold 2008

4
The brain and the senses

Small brains can do more than run at minimal levels of functionality and performance often exceeds human expectation. The avian brain offered particular challenges to evolution of its functional capacity for two reasons: a bird brain has to be small and well balanced to be aerodynamically viable and because, as a physical structure, any brain belonging to a vertebrate is nutritionally very costly. Full capacity (or functional ability), even within a given species, may be difficult to achieve or to maintain. If a brain does more than provide a coordinating system for basic survival skills, it is a luxury – and is possibly achieved at the expense of other mechanisms or means to maintain an organism.

From a human point of view, for a long time the argument was that a small brain can do far less than a large brain and the idea of innovativeness, decision-making, problem solving, let alone insight or the idea of remembering the past or planning for the future, were considered unachievable by brains as small as those of birds. There were two major and seemingly unrelated processes that contributed to a change of direction in avian research. One was the development of computers for home use and the other was an increased interest in vocal learning. In the 21st century, computers have developed very quickly and shown that memory storage is not always a matter of size but of the structural capacity and connectivity reducing the space needed in which an ever-increasing number of items can be stored, retrieved and cross-referenced. The second issue, vocal learning, is discussed in Chapter 10. For now, this chapter will consider only brain size as an issue and the brain's perceptual apparatus and how it is involved in sensory input. Research in this area will be related to magpies.

In 2001, several deceased magpies were donated to our laboratory (with relevant permissions) and this offered the first opportunity to weigh, measure and examine the magpie brain. Our main interest in the brain was initially for one specific reason. The neuroscientific literature on avian song had made the implied argument that only male songbirds developed song and did so largely as a reproductive strategy, i.e. to attract a female. Since my own collection of magpie song was rather large even at that time, it was clear that magpie females and males both sing. At the start of the 21st century, the scientific literature had reported on very few species in which song was known to be an ability found in females[1] and in the few species that were subjected to a neuroscientific examination it was stressed that there were sex differences in brain structure and the song nuclei in the brain and that the female brain was 'simpler' than the male brain.[1] Hence, the very first task was to examine juvenile and adult brains of male and female magpies and establish whether there were any differences. There were not. Developmentally, females were a little ahead of males but all the necessary infrastructure of a song control system was present in both males and females and, as importantly, also used by both males and females to produce equally complex song.[2]

There was a second reason for examining the brains of magpies closely. The human brain has a characteristic set of layers while the bird brain has a smooth surface (Fig. 4.1). Indeed, it was thought that the layered structure was a precondition for cognition and birds lacking this structure therefore allegedly lacked cognitive ability, an argument that had become increasingly untenable. Indeed, in 2004 an international consortium changed the name of all parts of the avian brain stressing more of the similarities with, than differences from, the human brain.[3]

The basic structure of the brain or the overall architecture of the brain, as Fig. 4.1 shows, is very similar in birds and mammals: forebrain, midbrain, hindbrain, an optic tectum and a cerebellum connected to the spinal cord. While the avian forebrain (now called the pallium) does not share the layered structure of the mammalian cortex, the structural similarity ought to be considered remarkable. It is likely to be an example of convergent evolution since the last common ancestor of birds and humans was as distant as probably 300 mya. Even more remarkable is the now known fact that these very different brains share many of the same functions, in particular those known to involve higher cognition.[4]

The brain of the magpie is noteworthy for another reason. Magpies have large brains relative to body size and a large Wulst area, a trait apparently shared by most corvids but not as pronounced in other songbirds. As Fig. 4.1 shows, in the magpie brain the Wulst is visible as a rounded protrusion (see arrow) and this is large for a songbird compared to the brain area. The Wulst (or hyperpallium) is especially well developed in raptors such as eagles, hawks and owls, all birds with broader binocular fields. Visual Wulst neurons may be selective for orientation and direction of

4 – The brain and the senses | 43

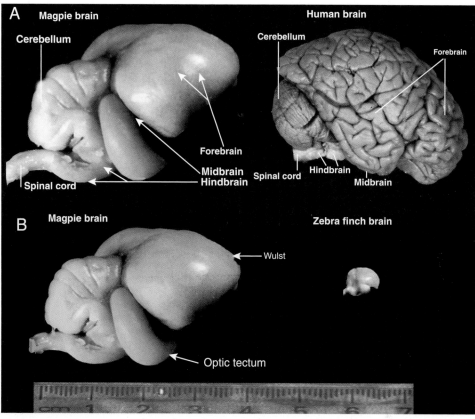

Fig. 4.1. Brain comparisons. A, not to scale, shows the right/lateral side of the magpie (left) and the human brain (right) with labels of major brain areas. B, for comparative purposes (scaled to size) again a lateral image of the magpie brain (left) and that of a zebra finch (right). Despite the drastically reduced size, the zebra finch brain has the same components and in the same order as those of the magpie brain. Note that most avian brain research has been undertaken on zebra finches.

movement.[5] There is also a portion of the Wulst that may process information of touch, in the case of birds relating to the feet (rather than the brain/hand coordination in humans). Raptors capture prey items with their feet. Hence, foot and eye coordination is of utmost importance to them.[6] Magpies do not use their feet for hunting but they may grip food with one foot and may thus require some foot coordination. I will come back to this point below, under brain lateralisation.

Brain size

Generally speaking, brain size is proportionate to body size; so when scientists started hunting for physical evidence of smart brains, there was little point of comparing brain weights/volumes of different species as a single measure (although that was seen as good enough even up to the 19th century). There is generally a correlation (scaling) between body and brain size – i.e. the size of each goes up in parallel and the largest birds in body size, such as the cassowary or emu, also tend to have the largest brains. By correlating body size and brain size this biological scaling rule is accounted for. As a consequence, evidence was sought to identify brain sizes that were larger than would be expected for a given body size. It has taken years in Australia but, finally, a ground-breaking study reported brain volumes of Australian avian species and relevant body measurements so that these could be correlated[7] (see also the detailed appendix for all Australian birds).[8] This measurement has been used worldwide and seems to hold as a rough indicator of greater cognitive ability in species whose brain volume is larger than the expected mean for its bodyweight class.

Fig. 4.2 shows the relationship between body and brain weight for all Australian songbirds as a single dataset. The centre line provides the mean relationship (while the upper and lower faint lines are the standard deviations). The dots are the data points for individual species. The species above the mean and especially species at the upper end and furthest from this midline are of particular interest because their position on the graph means that their brains are larger than those in the cohort and adjusted for their body size. The dashed line inserted into the dataset represents the family of magpies, butcherbirds, currawongs and

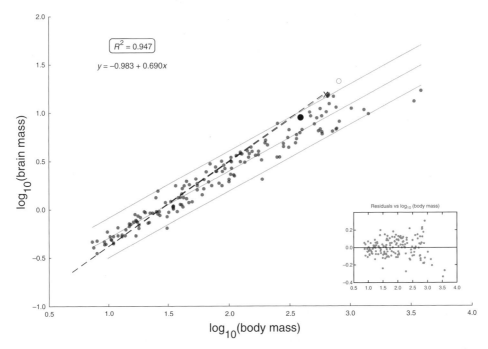

Fig. 4.2. Brain weight (mass) versus bodyweight for Australian songbirds. Centre line: mean, above and below lines standard deviations, stippled line results for artamids. Large black dot represents position of the Australian magpie. Black cross indicates position of raven and hollow circle, for comparison, the palm cockatoo (the latter not a songbird, of course).

woodswallows. As can be seen clearly, in this family (except for a few woodswallows), brain weight to bodyweight is greater than the mean for other species, i.e. they have larger brains than would be expected.

The large black dot in the upper section represents the magpie, on the same trajectory as the raven (marked x). Among songbirds, Australian corvids and Australian magpies as well as some butcherbirds (grey, pied and black), the pied currawong, the Australian raven and the Torresian crow are royalty as far as brain size, relative to bodyweight, is concerned.

In Fig. 4.2, a cross shows the position of the common raven of the Northern Hemisphere, very much in the same position as the Australian raven. A small hollow circle above indicates the position of the palm cockatoo in this ranking exercise. The Australian sulphur-crested cockatoo and its close allies and the New Zealand kea, as well as some of the South American macaws, top the list of brain sizes relative to body sizes in birds in the world.

Within their own group (Fig. 4.3), of course, magpies are not so special because most species belonging to the artamid family are very similar and hence cluster around the mean. Note that the pied and great butcherbirds, however, are shown as the highest ranking (larger brain relative to bodyweight than even the magpie).

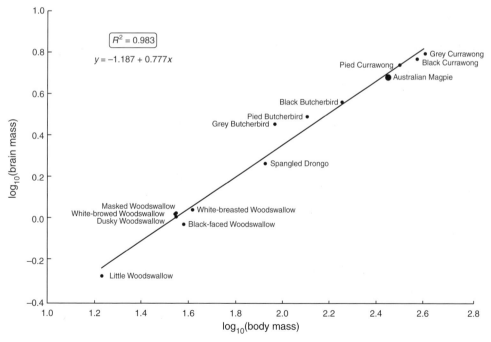

Fig. 4.3. Ratios between body and brain weight for Artamidae. Note that these ratios are identical for the Australian magpie, the pied currawong and the New Guinean black butcherbird. The inclusion of the spangled drongo here as a corvoid bird outside the artamid family but in the weight range of some butcherbirds is well below the ratio for butcherbirds. The grey butcherbird, for instance, has a much larger brain mass to body mass than the drongo despite a very similar body mass.

Other remarkable songbird groups (not shown here) are the bowerbirds, some fantails and honeyeaters, and even the zebra finches also punch well above their weight in brain size and pulling up the mean to a very high standard (for instance, waders/shorebirds and some raptors together have a much lower mean but they are not included here). Another group with substantially larger brain volume adjusted for size are the cockatoos, some of which, on this simple correlational measure, are well above the much-celebrated raven as the 'most intelligent bird on the planet',[9] an exuberant statement that may reflect some geopolitical pride of the Northern Hemisphere rather than biological evidence.

Having established these facts, the question is how one tests whether these details are of any consequence and really mean that they indicate special abilities. It remains to be seen whether more brain volume might actually translate into function and enable cognitive feats – and which ones. For neuroethologists, brain size is only the roughest measure as comparative work on avian species has progressed well beyond this now. Another question to ask is how brain volume is achieved. We know that birds that have a special need to remember locations tend to have a larger hippocampus than other songbirds. Birds with large song repertoires may have a larger song control system. Therefore, it is known that the brain can allocate additional space for functions that are of paramount importance (called exaptation – using existing neurons for new functions). We already know that the number of neurons can vary in different brains, sometimes even within a single species and throughout the lifetime of a single individual.

There is a method that was first proposed in 2005 that can actually measure neuron numbers and densities in the brain.[10] It needs a special piece of apparatus, an Isotropic Fractionator, capable of measuring the number of neurons in a brain.[11] Given the right equipment and expertise this can now be done, as has been shown by a team at the Charles University of Prague.[12] Importantly, they not only measured the overall number of neurons but that of the different parts of the brain separately (be this the cerebellum or the midbrain or forebrain, as per Fig. 4.1), meaning that the number of neurons in the forebrain, where cognitive processes occur, can be identified and compared with the same areas in various species.

This revealed a very important and perhaps surprising principle of scaling. While body size tends to co-vary with brain size, brain size does not necessarily co-vary with neuron numbers. To explain: a bird with 10 times the brain volume of another bird may only have twice the number of neurons (Fig. 4.4). Expressed in converse order: the smaller a bird's skull gets, the fewer neurons it may lose!

Since we have no measure for the Australian magpie of actual neuron density, the rook (a European corvid) has been used instead in Fig. 4.4 because the species is of similar weight range as the magpie. Note that Fig. 4.4B shows that for all pallial masses, birds have significantly more neurons per unit weight, i.e. they have higher neuronal density/are more densely packed. This is interesting to note, suggesting that there are good biological foundations for investigating cognitive

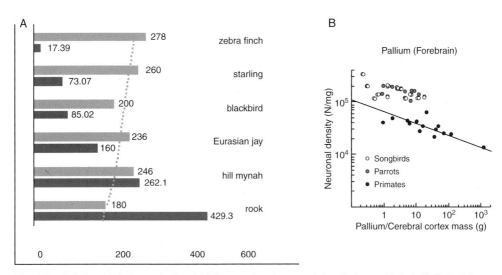

Fig. 4.4. Relationship between bodyweight (in grams) and neuronal density in songbirds (mg). A, Dark bars indicate bodyweight (number refers to average bodyweight). Light grey bar above refers to neuronal density (ratio: brain mass to neurons in mg). An example: a zebra finch of a bodyweight of barely 17 g has a neuronal density of 278 while a rook of nearly 25 times the bodyweight of a zebra finch has a lower neuronal density of just 180. B, The principle is maintained even when including mammals, here primates. Neuronal density relative to body size, even cortex mass. Source: B, Olkowicz et al.[12]

abilities in birds. One may even hypothesise that bird behaviour, certainly of the artamid family, may be on a par with primates and, in some cases, even with great apes.[13] The findings shown in Fig. 4.4B made news around the world[12] because now the overall size difference between birds and mammals (especially primates) showed the difference in functional brain capacity was likely not to be as great as had been assumed. Indeed, birds, and especially the Corvoidea to whom the magpie belongs, seem to have been endowed with cognitive capacities well outside their weight range. We are now in a position to be able to match some of these biological findings about the brain with results on magpie behaviour established in the field. Indeed, some of the magpie's remarkable aptitudes will be discussed throughout the remainder of the book.

Lateralisation

The brain of vertebrates consists of two halves, the left and right side. The question that has been of longstanding scientific interest is whether the left and right sides of the brain are duplications in the sense of function, or carry out different functions that increase the efficiency of the brain.

Hemispheric lateralisation of the brain, fully established as being common to all vertebrates, has now also been studied in some invertebrates, such as bees, revealing that even invertebrates have some hemispheric specialisations. The best-known example for brain lateralisation in humans is whether someone is left or right-handed, but there are many forms of laterality which can be measured behaviourally.

The eyes in most birds are placed laterally on either side of the head instead of frontally, as is the case in humans, most primates and predators generally. One of the obvious consequences is that the binocular field in birds is thus relatively small, while nearly 300° of all viewing angles are naturally monocular. This has one advantage in that so-called 'blind areas' of vision tend to be much smaller in birds than in species with frontally placed eyes (see diagram insert in Fig. 4.5 showing the specific viewing range in magpies).

But true for humans, other mammals and birds alike is that input from one eye goes to the contralateral side of the brain. In other words, input from the right eye goes to the left hemisphere (Fig. 4.6 below). For vision and hearing (but not of olfaction), sensory perception is thus processed opposite to its input.

Studying laterality can provide important clues how and where the brain processes sensory information. It is particularly pertinent in songbirds and many non-songbird species since their eyes, as in some mammals (for example, horses), are positioned laterally in such a way that one eye will often see an entirely different image from the other eye. Viewing with one eye and turning the head to the right or left deliberately so that a particular eye is attentionally engaged in the viewing, tells us which brain hemisphere is being used.

To explain this further: the magpie looking at the camera in Fig. 4.5 (and examining it closely) literally only sees the camera with one eye (right eye in this example) while the other eye receives an entirely different image.

As we now know from a large number of studies, largely of the domestic chicken and the pigeon,[14,15] the avian brain is specialised to process perceptual inputs differently in each hemisphere and to control different motor functions by each hemisphere (i.e. by each side of the forebrain or pallium).

In fact, one side may modulate or even suppress responses controlled by the other side. The information gleaned simultaneously from the other eye may be suppressed but may still compute and be capable of noticing and attending to a predator overhead, for instance. It is thus possible for a bird to forage, focusing on the ground with one eye, at the same time maintaining vigilance for danger overhead with the other.[16,17] Several studies have shown that a novel stimulus is more often viewed monocularly with the left eye rather than with the right eye.[18,19]

Therefore, an important facet of the vertebrate brain, birds included, is that both sides of the brain carry out different functions even though there may be no differences in gross structure between the left and right sides.[20]

To argue that lateralisation is an advantage because it may expand options to respond to the world is only one element of explaining brain lateralisation. Another important element is that some functions can be utterly incompatible and

Fig. 4.5. Monocular and binocular vision in magpies. Each eye of the magpie has 143°–149° of monocular field and the binocular field is only 34°. Head turning to view stimuli is seen commonly in the activities of foraging and monitoring for predators.

require different types of processing (responses such as approaching or fleeing) and by separating them, allows for very measured or fine-tuned responses.[21,22]

To date, few studies have investigated lateralisation of avian species in their natural habitat and even fewer for eye or ear preference, but the number is growing (for example, see endnote 27).

A way to measure functional differences of the hemispheres is to record eye preference for a specific item or situation. Lateralisation of visual processing is relatively easy to measure in wild birds with eyes positioned laterally. They turn the head to view stimuli at a distance using the lateral, monocular field of vision. In so doing they ensure that the hemisphere contralateral to the eye being used is primarily processing the incoming visual information. Eye, and hemisphere, preferences can be determined quite easily by measuring the angle of the head adopted by the bird to view a stimulus. We measured it in the magpie and found a

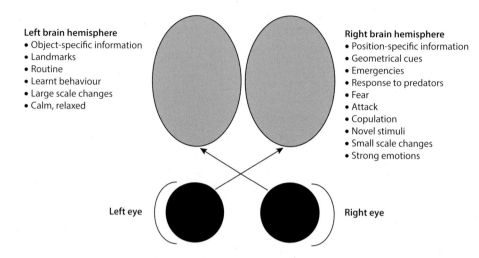

Fig. 4.6. Functional aspects of the left and right sides of the brain. Note that visual input to one eye is processed by the opposite hemisphere.

range of ~28°–34° maximum of binocular vision and most of such vision was limited to close range (Fig. 4.5).

In summary, avian brains have been found to attend to different stimuli and in different ways, as is summarised in Fig. 4.6, showing the two halves of the brain and some of the functions that have been reported.

In an experiment with free-ranging magpies, we supplied the magpies with food by purposely throwing pieces of mincemeat in their direction and then scoring which eye they last used before taking and consuming it. Of 155 scores, 97% were left-eye dominant, meaning they involved left-eye viewing the moving target before food retrieval.[17] The results of this field study are consistent with preferred use of the left hemisphere and right eye in control of feeding responses – as has also been shown in other species, first discovered in the chicken[23] and later also confirmed in other birds, including Australian species such as the zebra finch[24] and the budgerigar.[25]

Attention to predators

Field studies of behavioural laterality in birds are still relatively rare, but the few undertaken so far have shown that laterality may also play a role in vigilance behaviour and may guide their interaction with predators.

We wanted to know how magpies respond to predators, what strategies they use, how they process the information and whether such processing was reflected in eye use in wild magpies faced with predators they encounter in their own territories. We scored eye use when predators appeared overhead, i.e. by natural

Fig. 4.7. Looking up for danger with the left eye (cf. Rogers and Kaplan).[17] Photo: Andy McLemore/Flickr, CC BY-SA 2.0.

observation in the field.[17] The typical investigative viewing posture by magpies is shown in Fig. 4.7, an exaggerated postural change of the head, stretched neck, sometimes swaying slightly in a semi-rotational way, until the bird has fixated the object and has decided on a strategy.

Reports of magpies responding with alarm calling to a wedge-tailed eagle shape high in the sky, would suggest that their vision is exceptional. The anecdotal evidence, together with the size of the Wulst, would lead one to conclude that the magpie's long-distance vision is excellent and its acuity is superior to that of the average human.

By setting up experiments in the field presenting magpies with taxidermic models of predators (Fig. 4.8) known to overlap with specific magpie territories in which we conducted the experiments, we were able to examine closely how magpies approached and viewed the specific predators.[26] Using such models, we were able to test for behavioural expressions of hemispheric specialisation in detail and in several sensory manifestations.[27]

Magpies walked or flew around the models and we were able to distinguish between clear events of inspection, approach, withdrawal and decision-making time, scoring the use of eyes while they assessed the model. Since this has been published,[28] suffice it to say here that we found that magpies consistently used

Fig. 4.8. Taxidermic models used in studying responses to predators and hemispheric specialisation. A, wedge-tailed eagle; B, lace monitor. We also used brown goshawks and little eagles, all know to occur within territorial boundaries (for more detail see Koboroff *et al.*).[28]

the right eye/left hemisphere before an approach and the left eye/right hemisphere before withdrawal from the model. We found that the right hemisphere controls a suite of anti-predator strategies based largely on scoring the use of the birds' eyes).[28]

In another series of experiments, magpies were tested in order to establish whether they preferentially view things other than predators with one eye rather than another. Magpies approached from behind by a human emitted more alarm calls when they turned their head to view the human with their left eye while juveniles begged for food on the right side of parents (as is shown in Fig. 4.9B) more often than on the left side.[29] These, and other examples discussed later, indicate that brain asymmetry is an important part of being able to navigate the world successfully. As is now more apparent than before, cognitive ability may be linked to the strength of lateralisation of the brain.

Auditory perception and lateralisation

Without a doubt, vision and audition are the most well-developed sensory abilities both in birds and in humans, and they are often used in conjunction: for example, there is plenty of evidence that learning is particularly effective and often more powerful when vision and audition are coupled.[27,30] Indeed, in the natural world, stimuli are often multimodal, meaning perception of something in the environment may have visual and auditory or visual and olfactory aspects and it may be important to integrate both for a full appreciation and ability to respond. In both vision and hearing birds tend to perform a little better than humans.

We tested the hearing range of magpies and although our sample was small (only three birds) the results for all three birds were rather uniform. We established that audible sounds perceived by magpies may range from 0.2 kHz to 7 kHz, requiring higher sound pressure levels (SPL) for the very low frequencies (below 1 kHz), as well as for sounds above 5 kHz.

Magpies feed exclusively on the ground, and they walk, putting one foot before another, while foraging, sometimes referred to as 'walk-foraging'.[31] Their ground feeding habits make them easy to watch in open fields. However, importantly, they are also extractive foragers, that is, they take food from below the ground that is not visible and does not always leave visual cues on the surface as worms may sometimes do. I established elsewhere[27] that, in many areas of Australia, magpies forage for scarab larvae, often known as Christmas beetles, a very destructive pest that causes damage to pastures during the larval stage, as well as to trees on emerging from the ground as flying beetles. Earlier research[32] had shown that scarab larvae create no visual surface cues and that magpies are capable of detecting them by the sound they emit alone. Juveniles are rarely successful and usually require at least five months post fledging for them to have even moderate success in finding and extracting scarab larvae from the soil and one can surmise from this long learning period that it is not an easy skill to acquire.[17,27] Extractive foraging in primates has always been regarded as a cognitively advanced ability precisely because the food cannot be seen but its presence assumed provided the correct steps are taken.

The question was whether magpies had a preferred side/ear of listening. The observations were conducted near Armidale, New South Wales, on the Northern Tableland. Here, largely sheep-grazing country, at altitudes of ~1000 m, three species of scarab beetles were strongly represented.

Farmers love magpies and encourage them onto their properties because they are the most effective pest control of scarab larvae to date.[33] The larvae may pupate and emerge as beetles any time between November and March, i.e. larvae reach their full size at exactly the time when magpie offspring fledge (around September, sometimes earlier-depending on weather conditions) and make the greatest protein and food demands on the parent birds.

The steps of securing larvae from below the grassy surface, as described in detail elsewhere,[27] and summarised here, was almost identical in any magpies observed and in different territories: the foraging bird: (1) scanned the ground walking slowly; (2) then stopped and seemingly looked closely at the ground binocularly; (3) then, holding absolutely still; (4) in the last moment, turning the

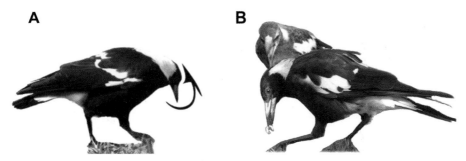

Fig. 4.9. Auditory monitoring for a scarab larva under the surface, characteristic head-tilt before jabbing and securing the larva, using the left ear, i.e. processed by the right hemisphere. A, the last minute swivel of the head to align the left ear with the ground before righting itself and delivering a strong jab into the ground to retrieve a larva. B, the impatient juvenile vocalising to the right of the female, waiting for food. The adult will first remove the hard head with the pincers before feeding the larva to the offspring (adapted from Kaplan).[27]

head so that the left side of the head/ear was close to the ground; (5) finally, the bird straightened up, then executed a powerful jab into the ground; (6) retrieving a large scarab larva from below the grassy surface. The important point here is that there was nothing to see on the surface so that the response could only have been brought about by listening to a sound. In all observed cases, before the bird actually jabbed its beak into the earth, it first swivelled its head so that the left ear was closer to the ground. It often had to deliberately change head position in order to gain proximity of the left ear to the ground (Fig. 4.9A, Plate 10).

These examples, along with others to come when discussing vocal performance in magpies, already show that the magpie displays several behaviours that indicate highly lateralised functions in vision and audition.

Should the argument be that the more highly lateralised a species is, the more likely it is that it may be cognitively complex? Indeed the evidence is mounting that in certain cases one is intertwined with the other.[34] If neural circuits have specialised to such a degree that they can carry out specific functions and in different hemispheres, then sensory input may also be more reliably and quickly assessed and accessed. Incidentally, corvids so far studied are known to be highly lateralised and their cognitive ability, ranking high among songbirds,[35] seems to be more than spuriously correlated.

Importantly, researchers[36] have shown that strong lateralisation creates a parallel working memory; in the case of pigeons for spatial location and food-related object cues, that also increases the information processing speed. Indeed, those largely interested in human brain activity, but often engaged in comparative studies with other vertebrates, have argued for some time that several cognitive abilities have been shown to heavily rely on lateralised processing in the brain,[37] the most widely investigated being language.[38,39]

Other cognitive domains that depend on lateralised processing include face and body perception,[40] spatial attention,[41] fine visual discrimination tasks,[37] motor skills,[42] complex problem solving,[34] specific spatial processing[43] and memory generally.[44]

The magpie certainly seems well endowed. The question is, of course, in what way and to what end the magpie uses its cognitive power, as other chapters will explore further.

Endnotes

[1] Odom *et al.* 2013
[2] Deng *et al.* 2001
[3] Reiner *et al.* 2004
[4] Jarvis *et al.* 2005
[5] Pettigrew and Konishi 1976
[6] Wild *et al.* 2008
[7] Franklin *et al.* 2014
[8] Kaplan 2015
[9] Jønsson *et al.* 2012
[10] Herculano-Houzel and Lent 2005
[11] Ngwenya *et al.* 2017
[12] Olkowicz *et al.* 2016
[13] Rogers and Kaplan 2004
[14] Rogers and Andrew 2002
[15] Rogers *et al.* 2013
[16] Rogers *et al.* 2004
[17] Rogers and Kaplan 2006
[18] McKenzie *et al.* 1998
[19] Vallortigara *et al.* 2001
[20] Rogers 2007
[21] Sherry and Schacter 1987
[22] Vallortigara 1992
[23] Rogers and Anson 1979
[24] Alonso 1998
[25] d'Antonio-Bertagnolli and Anderson 2018
[26] Koboroff and Kaplan 2006
[27] Kaplan 2017a
[28] Koboroff *et al.* 2008
[29] Hoffman *et al.* 2006
[30] Eales 1989
[31] Brown and Veltmann 1987
[32] Floyd and Woodland 1981
[33] Goodyer and Nicholas 2007
[34] Magat and Brown 2009
[35] Clary *et al.* 2014
[36] Prior and Güntürkün 2001
[37] Ocklenburg *et al.* 2014
[38] Corballis 2012
[39] Ocklenburg *et al.* 2013
[40] Thoma *et al.* 2014
[41] Duecker *et al.* 2013
[42] Arning *et al.* 2013
[43] Tommasi and Vallortigara 2004
[44] Habib *et al.* 2003

5
Diet and cognition in foraging

The ability to feed is one of the fundamental necessities of life. Many experiments on cognition in birds have centred around the question of how they feed and whether any of their foraging habits require any special skills or cognitive abilities to enable them to remain healthy and survive.

For instance, the question on when and where a food is retrieved has become a major topic of study particularly in nectar-feeding birds, such as honeyeaters, and a large range of fruit-eating (frugivorous) birds. These species must somehow know and remember when and where nectar can be obtained. This includes knowing from which plant or flower nectar is available and when such a food resource (in the case of nectar) would likely be replenished. In other words, they have to form a memory of what-where-when, called episodic memory, that tells a bird precisely what flowers it visited, where this had taken place and when it last fed from them.

In small birds, such memory can be a matter of life and death. The smaller an organism is, the higher its metabolic rate because its surface area relative to bodyweight is larger than in larger animals. It therefore loses heat more rapidly and requires a higher energy intake. The larger the animal, the slower metabolism becomes. Hence hummingbirds, among the smallest birds, need to constantly 'refill' or lose energy rapidly to a point when they could not even fly again to find another food source. Part of their efficiency derives from having developed the kind of memory that takes them to the right food source and, crucially, at the right time.[1] Birds that feed on fruit must remember where such fruit-producing plants are, when the fruit ripens and, in some cases, must recall colour and appearance of a fruit. Some fruits produce toxins that are at their highest level in unripe fruit. Currawongs, of the same family as magpies, fall into this latter category. It is usually not well known that currawongs are largely fruit eaters and, apart from bats, are of vital importance as seed dispersers of native fruit and, without the currawong's feeding habits and digestive system (leaving the seeds intact) the demise of hundreds of native plant species would follow.

For insectivores, as magpies are, one might well think that many of the rules that govern feeding techniques and memory associated with identifying food sources might not apply. But one would be wrong in assuming this. It is true that different cognitive mechanisms may apply concerning, for instance, the extent of spatial memory needed for localised foods versus memory for food items distributed over vast areas. It may be reflected in searching techniques and in switching from local memory about the location of food to movement-based strategies.[2]

On the other hand, exploration of new food sources may be required if the known source has been depleted or has simply not appeared where it had been found before. One of the major advantages for fruit/nectar feeding birds is that the potential food source remains in the same location. Birds that search for insects are dealing with food items that are either very mobile themselves (and can even fly), or they are hidden underground. When hidden underground, it requires either knowledge of physical tell-tale signs such as small mounds of soil left on the surface by worms, or different forms of memory that will enable a bird to find the food or enable it to crack open a shell to access the nutritious parts. In New Caledonian crows (also a corvid and part of Gondwanan stock), for instance, the use of a stick enables them to procure larvae hidden in small cavities of trees. The use of a small twig that the bird inserts into any small tree holes is not for dragging the larvae out by spearing them but, by prodding them, the larvae use their strong pincers to hold on to the stick to immobilise it and it is then that the crow can pull out the larvae still attached to the stick.[3] Whether underground or hidden in a tree, extraction of food in a context that makes the food entirely invisible is considered a special cognitive skill because it requires the forager to *imagine* a food source and search for it without any immediate reinforcement or sensory evidence of its existence.

Of similar, or possibly greater, importance to the cognitive domain is the ability of an animal to hide and retrieve food. Hiding food, referred to as caching, has become a very important sub-theme in research on higher cognition in birds because it involves memory of food out of sight and planning for the future when the food is retrieved. It is a behaviour that can be observed and experimentally tested.[4] Caching appears to be a trait of most corvids worldwide. It has been shown to exist in North American species,[5] in the corvids of Europe[6] and also in crows, currawongs, butcherbirds and magpies in Australia.[7] However, in Australian species, the observations have been collected in free-ranging birds rather than being tested in experiments. Hence, we do not have experimentally verified data whether magpies, or any Australian corvids, can rival the performance of scrub jays, for instance, in: a) recalling reliably the site of the cache, b) memorising the contents and c) memorising even the degradability of the cached food and hence retrieving it differentially – degradable food is retrieved sooner.[8] Manipulating food that does not lend itself to immediate consumption as well as exploration of novel food sources have also been important topics of analysis. These and others are a few methods that have been identified as being of specific interest in cognitive research.[9] More about this later in this chapter.

Dietary requirements

Magpie diet includes earthworms, larvae, beetles, grasshoppers and other insects and invertebrates (especially ants and spiders), small vertebrates (frogs, lizards, birds), carrion and grain.[10] An extensive investigation into the food choices of magpies discovered a diet of over 140 different items (see Table 5.1).[10] Magpies, like currawongs, will occasionally scavenge at carcasses and dead meat, paying particular attention to bone marrow.

Since cognitive behaviour in birds has become a topic of major interest worldwide, researchers have started looking at the dietary requirements of avian species more closely. As a first justification for doing so one can say that foraging is basic to survival. However, food is not evenly distributed nor is it always easy to spot, handle or extract and may require special skills or, depending where the bird finds itself, even new skills.

One way to classify food items is by the ease with which they can be seen and obtained. In the case of insectivores, as magpies are, items simple to spot on the surface and slow in movement may be at the lowest point of skill development (such as picking up snails, larvae and slow-moving beetles). Perhaps the next skill level is to detect well-camouflaged or fast-moving insects, such as millipedes and cockroaches. More difficult still are those insects that can either bite or are

Table 5.1. Summary of foods consumed by Australian magpies throughout Australia.

Plant matter	seeds, grains, dicotyledons, tubers, figs, walnuts, prickly pears
Invertebrates	earthworms, snails, millipedes, cockroaches, cockroach eggs, mantids, grasshoppers, crickets, phasmids, cicadas, antlions, beetles, beetle larvae, leaf beetles, weevils, ground weevils, moths, pupae, caterpillars, army worms and larvae, bees, ants (at least eight varieties), crustaceans, spiders and spider egg sacks, scorpions
Reptilian/amphibian	frogs, skinks, small dragons
Mammalian	mice, small rodents, meat scraps
Other	bird eggshells

Source: summarised from Barker and Vestjens[10]

exoskeletal, like millipedes and cockroaches, because hard or stinging parts may need to be removed before digestion. Gradually, the degree of difficulty can be raised simply by the speed or defences of the prey item. Finally, there are food items that have shells and require some form of manipulation and in the most difficult category are items that need to be extracted and are neither obvious nor plentiful. Some manipulation and extractive foraging has been classified as cognitively complex behaviour. If all the categories are mastered by a bird or mammal and occur specifically in the territory in which it is foraging, then this is called a complex foraging niche. A rainforest is probably the most complex foraging niche, although the open woodland that magpies preferentially occupy might still be called a relatively complex foraging niche.

Food innovations and extractive foraging

The magpie's range of food items may be large, but there is often substantial skill involved in acquiring it. Earthworms and larvae can usually be reached only by methods of extractive foraging, which requires special skills and adaptations. Magpies are able to locate these prey items by listening to the very slight sounds the prey make when moving. As explained before, a study had shown conclusively that magpies find scarab larvae by sound alone (and in some cases some vibratory cues) and not by olfactory or visual cues.[11] The researchers recorded the minute sounds of movements made by scarab larvae then, using very small speakers buried in the ground, they played back the recorded sounds. The magpies were able to detect the sounds, and locate and dig up the speakers!

The magpie's use of sound contrasts with that of other foragers, such as starlings and American robins. These birds find earthworms by visual cues, such as minute mounds of dirt heaped above the ground and left behind by the worms at the place where they had moved beneath the surface.[12]

The magpie's method of extractive foraging has to be learnt and fledgling magpies walk close to the adult, eventually linking the sound to the food (see also Chapter 8, the section 'Learning about food'). As was shown in the previous chapter, identification of scarab larvae occurred purely by listening and is an activity that is strongly lateralised.[13]

One cannot stress enough that acquisition of such food items as large scarab larvae is a feat of complex extraction, the kind that only some avian species (as cockatoos), some primates and predators (birds and mammals alike) have mastered, including foxes, or birds of prey that look into burrows, bears that find their quarry under rocks or in rotten trees, orang-utans that dig out plant bulbs from under the ground and chimpanzees that fish for termites in tree holes. The basic premise is the same in all cases: the food can usually not be seen or smelled and it is from past experience and exposure alone that any animal endowed with the appropriate cognitive ability will stop and attempt to acquire such a food item (and will have learned where to look).

Innovative feeding may also open up new food sources. For instance, currawongs and Torresian crows have been able to kill the poisonous cane toad and partly use as it as a novel food. Such stunning problem solving not only requires skill but insight. Researchers[14] have recently elaborated on the distinction between a complex skill niche versus a complex knowledge niche. Into a complex skill niche one may want to put foragers that have learned to break open hard shells, either by identifying a weak spot in the armour or by using tools or implements to break open a mussel, as white-winged choughs (*Corcorax melanorhamphos*) have been known to learn.[61] An example of a complex knowledge niche might be the ability to consume dangerous and poisonous insects, larvae and even amphibians by expertly removing or avoiding toxic parts of the animal.[9]

Magpies usually do not need to manipulate food in the way parrots do when eating fruit and nuts. One well executed jab is usually enough and the morsel of food is transported into the beak. As stated earlier, magpies have been known to tear strips of meat off carcasses if the carcass is exposed but such sightings are rare. In urban environments, dog bones that have been split or cracked tend to attract magpies. On finding a cracked bone a magpie may spend a considerable amount of time trying to extract the marrow. For this purpose, the magpie takes the bone in one foot, then, if the bone is not too heavy, tips it up so that the exposed side faces upwards (Fig. 5.1). Birds of prey also use feet to manoeuvre food but perching birds usually use their feet only for perching.

Eye/foot coordination in parrots has largely been studied in the context of food acquisition,[15,16] and particularly detailed in studies on owls, but in songbirds it has rarely featured because it usually does not exist. Songbirds pick up their food with the beak and feet are usually not involved. However, 'footedness' has been shown in some songbirds in conjunction with tool use.[17]

Magpies are very partial to bone marrow and will go to great lengths to extract it. Practically, there are real difficulties in marrow extraction – just using the beak usually does not lead to success. By provisioning magpies with pieces of large bones still containing marrow, it was possible to observe what they would do with them. Not unlike a parrot, the digits of the magpie foot can be used like a hand and magpies held the bone as if it were an ice-cream cone. Because of the weight of the bone, this position usually could not be maintained for more than a few seconds. The moment the magpies tried to peck at the tubular end of the bone, the entire bone moved as it did not have a stable base and could tip over easily. The birds were unable to retrieve marrow using this technique. However, the magpies did not give up. When this strategy did not work, the magpies used their left foot to steady the bone close to the end and then attempted to extract the bone marrow with their

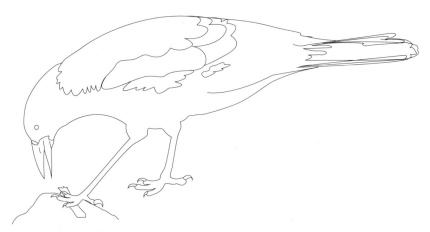

Fig. 5.1. Representation of a magpie manipulating a bone with the left foot in order to reach the bone marrow.

beak, bringing their head and beak down next to their foot but this could also fail if the bone pointed down too far.

Another technique that one magpie perfected unaided was to drag the bone to a small rock, mound or fallen branch and position the bone in such a way that it was angled up leaning against the rock or branch (Fig. 5.1). It then used its left foot to hold the bone at exactly that angle (preventing it from slipping to either side) and then the bird could easily extract the bone marrow, at least to the depth in the bone that the beak could reach. Once the other magpies watched and discovered what the adult male was doing, they quickly joined in and, by observation, soon learned the technique and secured their own bones with an exposed side for bone marrow. Since this was observed in one family only in which the adult male succeeded in the task, no conclusions can be drawn whether this is footedness and whether it is shared by all magpies and possibly even by butcherbirds and, if so, whether left-footedness is shared at the population level.

The successful extraction clearly also required some problem solving: dragging the bone to elevated ground and resting it upwards, the bird had found a workable solution to a problem. Artamids are generally accomplished in extractive foraging. Butcherbirds and currawongs will extract food from under the bark of trees, under logs and in crevices, and close to the soil surface. They will also store and dissect food at mid-storey level. Magpies, by contrast, stay on the ground for feeding and take to trees only for roosting and breeding. The only above-ground feeding by magpies occurs during breeding time when nestlings are being fed. After magpie nestlings fledge, they too are eventually brought to the ground and walk with a parent during foraging bouts (Plate 13).

Not all magpies may have access to food of the same nutritional value. Some researchers found that magpies without their own permanent territory, such as flocking birds, had lower bodyweight, higher mortality and were subject to disease at a higher rate than territorial magpies.[18] So far, however, the evidence is not conclusive that diet can entirely account for the difference in overall health between the flocking and territorial groups.[19]

Where and how magpies feed

One of the first ornithologists in Australia to pioneer the analysis of vertical feeding zones in native birds was Allen Keast.[20] He noted that the various species of thornbill had evolved important behavioural adaptations to their feeding strategies by subdividing vertical space. Some species were using only the upper canopy, some fed at mid-storey and others mostly on or near the ground.

All artamids use different strata for feeding, although they tend to overlap in their vertical terrains to some extent.[21] Woodswallows tend to catch insects on the wing and use the upper canopy. Woodswallows may also feed on the ground or occasionally in the mid-storey (looking for pollen and nectar) but generally they are not contesting the same space as magpies. Currawongs roam across all strata, from ground to upper canopy in search of food, while butcherbirds feed generally from the ground to mid-storey often seen pouncing on food on the ground but returning swiftly to a mid-storey position.

Magpies, currawongs, butcherbirds and woodswallows tolerate each other well within a given terrain, but usually by observing a foraging hierarchy. Magpies assume absolute first feeding rights on the ground while currawongs, as a semi nomadic species, stay back and never assert themselves when it comes to sightings of foods. Their interest in insects or protein of any kind tends to be limited to the breeding season (as food for their young), hence there is little to no competition for the majority of the year with insectivorous or grain-feeding species.

The magpie is one of the efficient generalists, although not as diverse as some. Magpies feed on a large variety of items (shown in Table 5.1 above) as long as these are found at or near ground level, usually either on or below the soil but never in the tree or in flight. The unusual ability for a songbird like the magpie to walk as humans do (not hop as butcherbirds and currawongs do or shuffle as the raven does) with possible large strides of left and right leg also determines its ground foraging strategy, referred to as 'walk-foraging'.[22] This is a walking behaviour that entails taking some alternate steps, pausing for a second or two before probing the ground in search of soil invertebrates and then resuming the walk. Food taken by magpies is usually stationary or is very slow moving, such as insects or lizards are, unlike kookaburras, for instance, that are sit-and-wait predators and tend to depend on the motion of the potential prey before they can detect it and strike. They usually observe the ground from above, a fair distance away, by sitting in a tree or on the wires of an electricity pole.

On very rare occasions, magpies have been seen giving chase to low-flying insects and leaping to catch them while on the ground. The latter might well be no more than a playful pastime exclusive to juveniles. We observed this during a locust plague in far western New South Wales. The juveniles walking along the ground stirred up locusts, which then took flight, and the magpies ran forward and took little leaps into the air, stretching out their necks and opening their beaks while running and leaping. Although the attempts at capture were only occasionally successful, it was noticeable that the juveniles tilted their heads and even part of the body in such a manner that they viewed the potential morsel with the left eye, as has been found to occur preferentially in other avian species when a moving target was detected.[23]

Magpies are diurnal, as are all artamids. However, magpies will continue to vocalise deep into the night, especially with clear skies,[24] and they may even fly about. There has been one observation so far of a magpie feeding at midnight, but as this was on a university campus and under bright artificial light conditions, we may surmise perhaps that, rather than being an occasional nocturnal feeder, magpies may require a similar level of light to that which allows human eyes to distinguish facets of their environment. There is no evidence otherwise that magpies can forage at night (under natural light conditions) although I have observed crepuscular foraging, especially during the breeding season, at light levels as low as 400 lx (Plate 13).

Caching of food

There is a difference between hoarding and caching. Hoarding, an extremely widespread practice among vertebrates, may consist of a wide range of things such as stealing of and then keeping ornaments, as materials for nesting or for no discernible reason at all. Some avian species are kleptomaniac, the most notorious probably the red kite (*Milvus milvus*) and the Eurasian or common magpie (*Pica pica*), not a relative of the Australian magpie. Red kite nest sites have been found to contain flags, handbags, magazine pages, tea towels, lottery tickets and socks – just as Shakespeare had warned in *The Winter's Tale* (Act IV, scene iii, line 23): 'When the kite builds, look to lesser linen' (see also RSPB[25]).

Food hoarding may be seasonally determined. There are larder and scatter hoarders. The former defines a clear place of depositing food, the latter indicates that individual items are hidden over a larger terrain. In some avian species of the Northern Hemisphere various forms of food hoarding are a matter of life and death since harsh winters with snow cover make food vanish seasonally. For instance, the fabled Clark's nutcracker (*Nucifraga columbiana*), has taken the cognitive task of remembering the whereabouts of up to 30 000 nuts under snow cover to an incredible extreme of spatial memory capacity. Remarkably, they manage to retrieve over 90% of them even over a span of several months and without the benefit of visual landmarks (since the nuts were deposited before snowfall and the ground remains covered under a deep cover of snow).[26,27] There are many other birds including corvids (ravens, jays) that use some form of caching.[28]

Caching is the process of hiding food. These habits of a wide number of species in the Northern Hemisphere and of higher latitudes than Australia, would not have attracted so much attention as a cognitive task had it not been for the way such food hiding exercises are managed. For instance, ravens and other corvids can distinguish between perishable and non-perishable food and will consistently retrieve perishable food first.[29] Even more telling is the fact that groups of birds may watch other birds caching in order to later steal the cache from the initial 'cacher'. And it gets yet more complicated when ravens, in particular, watch other ravens before they cache a food item – they look around and if there is another raven in visual contact, not even nearby, the caching raven will either deposit the item and then continue watching and only when the other onlooker has left, will retrieve the item and then hide it somewhere else or not hide the item as yet and wait until the bird has no onlookers. These very intentional acts presuppose that the caching corvids can assess the other's state of mind and its intentions, something that only humans were thought to be able to do.[30]

The Australian climate generally does not bring about the need to cache food and, in most cases, there is neither an abundance of food nor a predictable shortage. Magpies do not suffer the deprivations of Northern Hemispheric winters where caching may be an essential survival skill, as it is for the Clark's nutcracker.[31]

That caching behaviour exists at all in magpies is noteworthy and might be related to group hierarchy, territorial competition (see Chapter 6) and perhaps even inter-individual competition (see also Chapter 7). The behaviour has never been investigated systematically nor been tested experimentally. However, throughout Australia reports abound of people having seen magpies carry food in their beaks and running off with it to some hiding spot. Magpies cache food by pushing it under leaf litter, wedging it under logs or by inserting it into the soil. There have also been sightings of magpies placing food items among thick flowerbeds and scrub, hence above ground. Retrievals of these cached food items usually occurs within the hour by the same birds. Interestingly, there is some anecdotal evidence that caching magpies scan for other magpies while they are about to cache an item and, in some cases that I have observed during field work, they relocate the morsel if the caching process had been watched by another magpie. This suggests that there may be food competition among magpies and another magpie watching the caching might raid that caching site, as has been observed in ravens.[32] In such cases, the same conclusions would need to be drawn that magpies are aware of the other bird's state of mind and have some ability to predict future action.

There are two further variants in the magpie's behaviour of carrying away food that I have not found described in any species.

Fig. 5.2. Juvenile magpie carrying an arm and a leg of a sock monkey, a toy that the pet dog had first thoroughly demolished. Equally, magpies may carry food as a play object when they are well fed.

Magpies may carry food in their beaks as an expression of play behaviour, i.e. to taunt a conspecific, and such bouts occur very frequently, not usually with food, but with any number of objects (Fig. 5.2). Another variant occurs in adult magpie pairs without a territory.

There are cases, more of this in the next chapter, of bonded adult magpie pairs without a territory. They may get a little area in someone else's territory to rest and a daily pass within yet someone else's territory to snatch a drink. I have seen such pairs in several geographical locations. The current 'homeless' pair has made one of the large palm pots on the balcony of our house their safe home base. They come and assess their pot every day, and the purpose of the visit is to check on their deposits and safeguard their cache. Indeed, since they have used this pot for storage for over a year now, it is probably best described as a larder. Some food items are firmly packed in between the separating stems of the palm, some are hidden under pebbles and, now and again, they may actually bury an item, such as a snail.

Caching of food is a worldwide phenomenon known in ~200 species, including largely rodents and some mammals and birds. Over a third of them are long-term cachers, meaning that their caches are seasonal and consist of what we call 'non-perishable' foods only. This does not apply in Australia because of latitude and climate. Hence, any caches tend to be short term and may mostly consist of perishable items, including even meat, as we know of Australian little ravens[33] and forest ravens,[34] butcherbirds and some shrikes and, as shown, occasionally of magpies. Rollinson[35] who had observed caching behaviour in Australian magpies, argued that food caching behaviour in magpies may well be a response to the general backyard feeding of magpies, a point that had also been made by Thomas[36] and may be a reflection of regular interaction between humans and birds that receive gifts of food at times of food abundance.

The matter of caching has recently caught world attention not so much for its cognitive interest but for the consequences of climate change on caching. The reasons are simple and disconcerting. Temperature changes and moisture are likely important factors influencing even seemingly non-perishable seeds. Once a food item is cached, it is exposed to environmental conditions that can either maintain or degrade its quality over time and that may spell disaster for some species.[37]

Dunking and other feeding innovations

Dunking, the act of placing food in water, seems a widespread behaviour that has been documented in avian species worldwide including corvids, sea- and shorebirds.[38]

Dunking may occur for very different reasons.[39] The deferment of gratification (not eating the bread crumbs given but waiting for something better), is one reason why dunking has become an interesting behaviour in the literature. In a few species (such as ravens and herons) a morsel of bread is taken but not consumed and instead moved to a waterway and dropped close to shore where it eventually attracts small fish which are then procured by the bird. Dunking seems to be most commonly used for softening and even cleaning of food.

Dunking may also be a way of coping with otherwise unpalatable or toxic food. Recently, such a case has been observed in magpies[40] and the particular instant that the authors described has a more unusual context than most instances of dunking so far documented. The object of dunking was a mountain katydid, *Acripeza reticulata*, a bush cricket with very specific qualities. Its marking is cryptic with a grey patterned spidery bony appearance which seems covered in fine hairs. However, when cornered it can use a rather dramatic defence, a startle display (called a deimatic display) by suddenly spreading its wings and exposing an abdomen of bright red broad stripes, lined with small turquoise beads and subdivided by black lines. These are clearly colours that either mimic or signal danger. It is thought that the katydid is toxic to birds or at least the fluid it exudes from its abdomen is unpalatable or toxic.[41]

Fig. 5.3. Magpies catching and eating honey bees. (Source: Colin and Suzie, Noggerup, WA)

The magpie carried the katydid for short distances, interspersed with repeated bouts of vigorous wiping (slow, repeated dragging of the item on the ground) or thrashing (vigorous beating of the item from side to side on the ground). The adult eventually dropped the cricket but a juvenile picked it up, also dunked it and then consumed it. One wonders whether the adult knew there were risks or even that the exudate had to be eliminated first before it was safe to be consumed. Of course, this raises more questions but it is noteworthy that the adult treated this katydid with caution.

Another feeding innovation by magpies has recently been discovered by Colin and Suzie Fairclough from Noggerup, south of Perth, Western Australia, that so far, I believe, has not been described in the literature. They noticed that magpies were searching in the bushes of native fuchsias, not at ground but at magpie eye level. The fuchsias produce beautiful long and tubular bells in bright red. They obviously also produce nectar because these flowers are regularly frequented by introduced honey bees. The bells are so long that the bee has to crawl completely inside the flower and only a little of its abdomen remains visible. In that position, they are quite vulnerable and incapable of responding to a predator by flight. The magpies have learned to wait until the bee is completely enveloped by the flower and then catch it by the abdomen, slowly move it to the ground where they wipe off the sting and then consume it (Fig. 5.3).

Efficiency in foraging

One of the most erroneous assumptions about animals generally, is that finding food is hard-wired and any bird will find its food eventually. While views and insights might have changed, they probably have not changed enough to date to accommodate important learning steps prior to release of captive birds. Quite a few people in Australia raise injured or orphaned birds and eventually release them back into the wild.

Let us assume that a magpie was raised on a slurry (specialised for insectivores) mixed in with mince. There obviously is no such food lying about in the yard. Putting mealworms in dishes (a good nutritious food) is equally no preparation for the wild – mealworms do not litter the ground and foods usually do not come on a platter. Birds that have been raised in this manner with no other avenue to explore will usually die within a month or two post release unless they have been exposed to actual food items and these have been presented on a background in which such food is likely to be found. Food identification, discrimination and technique of handling food are learned behaviours – the only genetic help the birds get is that, predictably, they can learn and acquire the knowledge and the skills. If they are on their own, they do not have the advantage of watching adults and thus cannot learn by observation.

Trial and error learning is of little value when the bird cannot even guess what it can trial.

To give young birds a good start it is worth remembering that juvenile magpies have an extended learning period of several months post-fledging in which time they accompany a parent bird on the ground and increasingly look and check how the food was gathered, what it is and how it is handled and consumed. Since there are over 140 different items of different colour, shape, range of difficulty and behaviour, this is quite a syllabus. Youngsters also have to distinguish between edible and inedible items, of necessity increasing the knowledge base even further.

The actual learning part for magpies, as for many other birds, is to recognise a motionless shape in among leaf litter, grass and stony or pebbled ground and detect

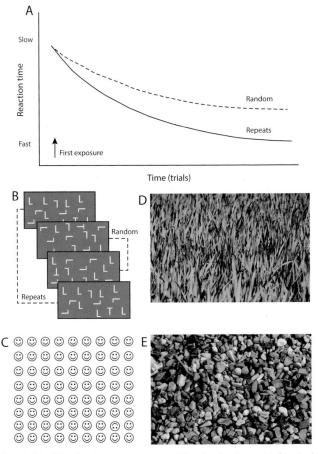

Fig. 5.4. It takes time to identify a food source among a distracting background. A, a typical response curve showing the length of time it takes to identify a specific shape amid similar ones. When an object is in the same place repeatedly reaction time is faster; reaction time is slower when the object is in a random position. B, C. Typical tests given to humans to test the reaction time of finding the odd one out (in B, find the letter T; in C, find the upside down face). Human tests are made simple by providing a clear background. Magpies, however, face backgrounds that are chaotic and powerful distractors (D, grass; E, pebbles) in the search for prey.

it as something edible. Knowledge of edible items has to go hand in hand with discriminating these items from others against a busy background and determining their edibility.

Birds need to visually cue onto the object and then quickly take action or most such items will simply slide, skip or fly away. Visual cueing is not such a simple task. Tests of recognising the odd man out (Fig. 5.4C) have become part of human intelligence tests. A second challenge is whether, once recognised, the object is worth tackling. A third component, equally important, is the speed with which action is taken, once a decision has been made. A typical test given to children is represented in Fig. 5.4B. Here, four cards are provided, all carrying the letter 'L' sideways. However, one of these letters is actually a 'T' and the subject is required to identify it out of all the others. Searching time can be quite extensive for the first trial.

If the second trial is a repeat of the first card (the lowest one in the A frame), then reaction time plummets and the 'T' is identified quickly. If the second card is shown instead (random distribution), reaction time increases again but not to the level of the first naive search. The reason is that the subject has learned at least which shape to look for and that, in itself, might improve reaction time. If the second random card shown (i.e. the letter 'T') is again in a different position to all previous examples, reaction time will be shorter even though it is random because the subject now knows the shape of the item to look for and that it will be somewhere on the board, i.e. confidence of its presence makes a search usually more methodical. If we now switch back to the ground-feeding juvenile magpie: the bird is bombarded with a series of random patterns and shapes and from the parent bird it gradually learns that among all this distractive patterning edible food is to be found. The more objects there are to learn about, the greater the demands on memory and thus cognition. None of these processes are very fast and, at first, not only inefficient but, at the beginning of the young magpie's learning, more often unsuccessful than not.

When speaking about foraging efficiency, there are at least four variables that are usually considered in research:[42] (1) the time it takes to find and process food; (2) metabolic costs accrued for each foraging activity (foraging, processing, etc.); (3) the passage of time for one item to be processed and successfully dispatched and, (4) how nutritious the item might be and whether it warrants the energy expense. As a formula, the simplest form is:

$$\text{Profitability of prey} = \frac{\text{Energy gain per prey item} - \text{Energy cost to acquire prey}}{\text{Time taken to acquire prey item}}$$

The end result of a foraging activity should be, of course, that there is an energy gain overall for each prey item secured. However, if it takes a long time to complete one transaction, for little energy gain, the bird might still be deficient in its energy uptake even if the bird has foraged for hours.

It seems that magpies are extremely efficient foragers judging by the number of hours they spend roosting, preening, socialising, napping and playing. Indeed, outside the breeding season, we kept regular records of the birds' time involvement in direct foraging acts and realised that magpies spend far fewer hours working to satisfy basic needs than their human counterparts!

These actions and decisions do not just have to be learned by juveniles, they remain important principles throughout a bird's life. Magpies live mostly in well-defined territories. One of the first tasks that the owners must learn and remember is the distribution of food in their environment if they are to make the appropriate choices on a daily feeding schedule. They might also have to remember when certain species might be abundant in one section of their territory and at what times of year, and how to access them.

Ultimately, how well a bird has learned to feed will determine its overall ability to survive, to breed and successfully raise young (fitness). That is by no means guaranteed as the records show that, in any given year, only 6–14% of all extant magpies in an area and in a given year may breed successfully (of this later). Therefore, there may be substantial differences between individual success, weight maintenance, longevity and ability to reproduce.

The social aspects of foraging

So far, some of the key elements of niche and foraging complexity have been outlined, showing that the diet breadth and the use of extractive foraging and feeding innovations truly qualify the magpie as belonging into the group of animals (such as higher primates) with complex foraging niches that have the ingredients of requiring both complex skills and complex knowledge. Hand in hand with species in such environments are offspring needs characterised by long learning periods. These are two of the hallmarks for all animals so far studied that have been found to have large brain and remarkable cognitive abilities.[43]

Another facet that goes hand in hand with protracted development, is the sustained support of adults. Magpies are a social species with stable group membership over several years[44] and, for the core members, some of these associations may last a lifetime.

Usually, when we speak of cooperative species, this refers to cooperation from helpers at the nest site, helping to feed the new brood. (This will be discussed in later chapters.) In some avian species, in particular the magpie, but also in noisy miners, apostlebirds, grey-crowned babblers and many more, this cooperation is extended well beyond the helping-at-the-nest routine. Magpies form pairs or families that work as teams and the best teams also last considerably longer than groups with less cooperative members. Magpies do not hunt cooperatively as do Harris's hawks (*Parabuteo unicinctus*)[45] and sometimes black kites (*Milvus*

Fig. 5.5. A pair of magpies setting out on their foraging activities for the day with obvious agreement where the foraging is to take place (the birds had just landed) and they tended to be in that same area at approximately the same time each morning.

migrans),[9] nor do they necessarily raise their offspring cooperatively but they will always, without fail, defend their territory together.

Interestingly, they also often forage together. Even if the magpies look as though they are singly foraging, at least a second one will be nearby. Indeed, pairs often set out on the day together (Fig. 5.5). Magpies belong to the few ground-feeders that forage socially.

In some groups, at particular times of the year, magpie groups (at least half the group but often the entire group) start foraging together in a formation that is reminiscent of search and rescue missions in human emergency situations when, for instance, a child has gone missing in a park or wilderness and volunteers comb the area in close proximity to each other. Such a formation is shown in Fig. 5.6.

So far, this behaviour has been observed only in the depth of winter on the New England Tableland (altitude 1000 m, latitude 30°32'S, 148°29'E) and late in spring. The strategy for the former may be related to the scarcity of food items and the kind of food items remaining during near frost conditions (underground worms and larvae are not accessible) and the few insects above ground tend to be fast-moving.

In spring, scarab larvae are incubating underground in large numbers on the Tableland's sheep grazing plains (Fig. 5.7). These larvae can destroy large areas of grazing surfaces and the magpies are very welcome to eradicate them or at least reduce the damage at a time when the larvae have high food intake and kill grass roots.

Fig. 5.6. Systematic foraging by a magpie group, aligned to increase foraging efficiency, spaced out evenly from each other (see also Plate 15).

Importantly, these larvae are larger than any of the typical magpie fare and of high nutritional value. If there is a way of very efficient foraging it is symbolised in the larvae, easily replacing 10 food items of the average fare. The problem is that finding them takes extra skill and knowledge and the very faint signals they emit in chewing sounds makes them all that much harder to locate. The more closely magpies are aligned and can scan an area, the less likely the larvae might escape their fate.

There is a third way in which magpies cooperate with each other in foraging activities. When a single adult finds a food source, the bird will issue a food call, aimed at attracting others of the group to the site. We know that many vertebrate species do this, from cockerels to capuchin monkeys. Sometimes, such signals may well be deceptive,[46] cockerels have been shown to give food calls so that the hens would approach and, when close enough, used the opportunity to mate with them.[47] Capuchin monkeys while under pressure to share may, when alone, first feast on the food source and only after they had their fill, then issue the food call usually showing some acting ability as if they had just found it![62]

After many years of recording their foraging habits, I would have to conclude that none of these formations of walk-foraging is accidental. They would seem to

Fig. 5.7. A, Scarab larva from a Sydney garden. B, Magpie with a grub it has extracted from the ground. Photos: (A) CSIRO; (B) Toby Hudson/Wikipedia, CC BY-SA 3.0.

represent a well-developed food searching strategy that, in many ways, is akin to sophisticated hunting strategies that some cetacean groups have developed, such as bubble-netting.[48,49] In each case, the purpose is to minimise escape/non-detection chances of the intended target. And, as shown in studies worldwide, such cognitive feats may be developed in one group and not another because a particular group may contain innovators. If innovations are successful, they get adopted by the entire group but not necessarily by a neighbouring group, as is known from studies of superb fairy-wrens (*Malurus cyaneus*).[50]

Mirville and colleagues[44,51] were particularly interested in cognitive aspects of magpie behaviour when studying foraging of Western Australian magpies. It is useful to examine these aspects in a geographical area where group size variation and large group sizes are more frequent than in the east of Australia. They confirmed that adults are more successful than juveniles in the solution of novel foraging tasks. While this is not altogether a surprising result, they also compared group sizes and their relationship to successful foraging. It is very noteworthy that, in their study, group size had an effect on the degree to which the novel food exploration was affected. More individuals interacted and solved the novel foraging task in larger groups than in smaller groups. Individuals in larger groups tend to be able to engage in more risk. This was also confirmed in a study of small passerines using blue and great tits in a foraging innovation set of experiments.[52]

Weather, climate and feeding

All of these variables of food consumption are under the influence of the climate, general weather conditions and rainfall, and the skill of an individual may really be tested. This is especially true in Australia, when droughts extend for months, the soil starts getting baked hard or when sudden heatwaves sweep a particular region.

And here it is necessary to restate a point that should be obvious to everyone: in a span of just two centuries, European settlers have radically changed the

Australian landscape. There has been dramatic deforestation – 70% of forests and woodlands have disappeared and most of the rainforests – two-thirds of arable land and half of all grazing land has been degraded or rendered useless,[53] rivers and wetlands have diminished or dried up and salinity continues to be such an urgent nationwide problem that each hour, so we are told, an area of the size of a football field is lost to salination.[54] The official *State of the Environment 2001* report found that Australia was the only developed nation among the top 10 land-clearing nations in the world.[55] Unfortunately, this picture has changed little, as the very disappointing and alarming statistics of the official *State of the Environment 2016* report have shown.[56]

More heat and drought cause major food supply disruption. Heat and drought have a direct effect on feeding behaviour because of the need for thermoregulation. We know that there is a direct relationship between the number of breaths per minute and ambient temperature. As Weathers and Schoenbaechler[57] had shown, budgerigars have about 50 breaths per min at a comfortable 30°C. Breathing rate begins to rise sharply at 40°C, doubling to 100 breaths per min, at 41°C again doubling to 200 breaths per min and then, importantly, every single degree of temperature increase seems to matter: at 42°C it is 200 breaths per min and at 46°C the number of breaths rises to 300 and then keeps increasing to 350 even without further increase in temperature. This depletes water in the body and may lead to hyperthermia or severe dehydration.

There are several methods of thermoregulation that birds use to cope with high temperature. The most obvious one is to seek shade (called microsite selection) that simply minimises exposure to the sun. Another one is wing-venting – the most efficient method for manageable heat because it does not lead to water evaporation. Other methods of thermoregulation are panting and gular fluttering, both methods that help in the short term but have the high cost of further water loss (Fig. 5.8).

Magpies, one would have thought, are in a much better position than other birds because of their size. We know that smaller birds are in danger of heat stress much sooner than large birds because they have less water storage in the body for evaporative cooling. They gain heat more rapidly when the ambient temperature is higher than their body temperature and do so for well-known physiological reasons as mentioned before. The smaller the organism, the higher the surface-to-volume ratios and the faster overheating can occur. We also have evidence that mass mortalities of birds and bats due to extreme heatwave events have occurred in Australia, cutting down thousands of budgerigars but also large birds such as Carnaby's black cockatoos (*Calyptorhynchus latirostris*).[58,59]

We are fortunate to have a study of the effects of heat on feeding behaviour in magpies. While birds, particularly in sudden heatwaves, tended to die by 42°C getting to catastrophic loss by ~48°C, effects on feeding and body condition

Fig. 5.8. Heat-stressed juvenile magpie photographed by author in the West MacDonnell Ranges, Central Australia, March 2015. The temperature was just reaching 40°C. The bird had stopped feeding, had identified an appropriate 'microsite', i.e. a tree with at least some shade and at a height that afforded a breeze. Within half an hour of sitting there, it had started wing-venting which can be seen in the image: the wings are slightly out allowing heat to escape from under the wing. It now also had its beak half open, panting. Fortunately, the temperature did not rise any higher that day but the bird was clearly heat-stressed and unable to move until the late afternoon.

alarmingly starts at much lower ambient temperatures, at least in magpies in Western Australia.

Edwards and colleagues,[60] who undertook this study on magpies, found that temperatures merely exceeding 27°C resulted in a significant decline in their foraging effort! This is a much lower cut-off point for feeding than one might have anticipated. Although magpies are able to shift foraging times to early mornings, this may not be enough in terms of their energy requirements if ambient heat remains or increases further over the day and afternoon.

Magpies thus trade-off foraging activity against heat dissipation. They display increased heat stress and reduce their foraging behaviours and if the heatwave occurred at the time of breeding, loss of bodyweight and body condition were substantial compared to non-breeding birds. The authors sound a warning when they say that these mechanisms are exacerbated by increasing temperatures, or by an increase in the number of heatwave events. In the event of such scenarios, they say, 'future body condition, longevity and reproductive success may all be affected',[60] an important reminder that birds as seemingly robust as magpies may not be physiologically capable of surviving through the heatwaves and temperature increases we have seen in the last 10 years. Magpies are as powerless against high temperatures as humans are and all the knowledge base in foraging does not help if they cannot feed.

Endnotes

1. Henderson et al. 2006
2. Sulikowski and Burke 2011
3. Hunt et al. 2006
4. Clayton and Dickinson 1998, 1999b
5. Balda and Kamil 1989
6. Buitron and Nuechterlein 1985
7. Chapman 1978
8. Emery and Clayton 2004
9. Kaplan 2015
10. Barker and Vestjens 1990
11. Floyd and Woodland 1981
12. Heppner 1965
13. Kaplan 2017a
14. Schuppli et al. 2016
15. Brown and Margat 2011
16. Randler et al. 2011
17. Noske 1985
18. Vestjens and Carrick 1974
19. Veltman and Hickson 1989
20. Robin 2001
21. Schodde and Mason 1999
22. Brown and Veltman 1987
23. Rogers and Kaplan 2006
24. Sanderson and Crouch 1993
25. RSPB 2009
26. Magnotti et al. 2015
27. Qadri et al. 2018
28. Sherry 2017
29. Clayton et al. 2005
30. Bugnyar and Kotrschal 2002
31. Vander Wall and Balda 1977
32. Bugnyar and Kotrschal 2002
33. Lewis 1978
34. Secomb 2005
35. Rollinson 2002
36. Thomas 2000
37. Sutton et al. 2016
38. Morand-Ferron et al. 2004
39. Lefebvre et al. 2016
40. Drinkwater et al. 2017
41. Umbers and Mappes 2015
42. Sinervo 1997/2006
43. Schuppli et al. 2016
44. Mirville et al. 2016
45. Bednarz 1988
46. Helgesen et al. 2013
47. Evans 1997
48. Wiley et al. 2011
49. Similä and Ugarte 1993
50. Langmore et al. 2012
51. Mirville 2013
52. Morand-Ferron et al. 2011
53. Lines 1991
54. Dovers 1994
55. Australian State of the Environment Report 2001
56. Australian State of the Environment Report 2016
57. Weathers and Schoenbaechler 1976
58. McKechnie and Wolf 2010
59. McKechnie et al. 2012
60. Edwards et al. 2015
61. Hobbs 1971
62. De Waal 2000

6

Managing a territory

The very concept of property or territory is not a human invention but forms the very basis of success for many species from insects (termites, ants, bees) and fishes, to mammals and birds. A territory can be everything – from providing all basic needs to the key physical area for raising offspring. Indeed, many species have to confine themselves to providing an appropriate shelter for young to be raised rather than claim an entire territory.

It is incorrect to believe that all magpies have territories. It is like human home ownership: while it may be desirable, only some achieve it and can retain it while many others miss out. The percentage of human home ownership is higher than that of magpie territory ownership but in both groups the percentage of the overall population claiming such a prized possession is declining and the properties are getting smaller. Such is the effect of expansion of the human population in areas that are also favoured by magpies. The number of homeless among magpies is also rising and some of the conflicts that have occurred near cities and towns between magpies and some less generous members of the human race, are usually a direct outcome of the removal of good habitat and roosting trees, the diversion or drying up of waterways and the covering of substrate with concrete and bitumen, as well as monocultural, often multinational agri-business undertakings that have not left a blade of grass, a shrub or tree standing as far as the eye can see – forming desert-like conditions for many species. I received an email from a man in Perth, claiming he had heard magpies cry most woefully when they watched powerlessly seeing their one remaining roosting and nesting tree in their territory fall. The writer said that the desperate cries of the magpies were permanently etched in his memory.

Magpies need territories to breed. Their preference is for a territory large enough and of a quality good enough to provide all water and food needs, enough structural support for roosting and an area with a low or manageable presence of potential predators. As in humans, the wish list for the ideal territory may be long but often not realisable and concessions have to be made – usually about the size and the nutrition available. Moreover, conditions can change from year to year and so can the size of the family that lives in it. Usually one speaks of stable and unstable territories.

Indeed, the number of individuals per group may ebb and flow from year to year as offspring leave while the parents stay and prepare for a new breeding season. Other resident pairs have helpers, often daughters from the previous year or years that are recruited for helping at the nest with feeding the new brood.

Having said this, in the case of magpies, the summary given above has been simplified as shall be shown later. Suffice it to say here that a description of territory is intertwined with group size and the group's social composition which has almost as many variations as exist in human society. Magpies live as couples, some males have affairs, some couples 'divorce', others stay together for life and some form rather large groups with adults and juveniles from previous years and possibly including even some unrelated adults.

The kinds of relationships I have been privileged to witness and follow reflected some of the extreme examples of this variability. Of the two stable pairs I knew, both in stable territories, one pair stayed together for 20 years and in that time produced a clutch of 2–3 offspring annually that left their natal territory before the next breeding season. After 20 years, they stopped breeding but continued living there. The other pair equally stayed together for 18 years, with similar success when, one day, the male did not return. The female stayed for a while longer but then disappeared.

In another territory, in a pair that was in the middle of raising two fledged youngsters, the male suddenly disappeared and the female was struggling on her own to provide food for the two fledged offspring. Very unexpectedly, the male that had disappeared flew in a week or so later with two females in tow and immediately began to assume authority over the territory and these two new females. The female parent was initially confined to a small area in the territory with her two fledged offspring but such a semi-truce only lasted for a few days. After continual harassment the female and her two offspring were forced to leave altogether, facing a very uncertain future, almost certainly condemning the fledglings to death. I have seen adult male magpies sneak away from their natal territory, on foot (even looking over their shoulders), at early morning hours to visit a female in the next territory, observations that have since been confirmed by detailed genetic studies.[1]

I have borne witness to multiple breeding events in the same tree or trees in close proximity to each other in *G. t. tibicen* subspecies (in New South Wales), observations that have also been made of the Western Australian subspecies *G. t. dorsalis*.[2] There were several cases in which the female of the breeding pair completely destroyed the eggs in the other nest as well as the nest itself. If there was ever an angry magpie she demonstrated her anger by loud and sharp vocalisations while ripping the nest apart. The female brooding on the then destroyed nest was a daughter from an earlier year. There was no obvious male partner attached to the daughter – reminiscent of situations and attitudes towards pregnant teenage girls in human society. The daughter was allowed to stay after motionlessly watching her mother's actions and then showing all the submissive gestures that the mother seemed to require even after the nest was destroyed. Another case of plural breeding seemed to involve the same female making two nests but this could not be confirmed, i.e. whether a daughter was tasked to brood the second nest.

I have seen eviction notices delivered by one group to another, with success and own offspring moving in, leaving the defeated group or pair without a home. And then there was the utterly unlikely story of four male magpies from different clutches that I had hand-raised together and who stayed together for life, after 10 years accepting one female into their group who then produced two offspring per year every year that were then raised jointly by all four males while the territory was extremely well defended against many incursions and birds of prey.

All these examples may or may not represent unusual events and overall may account for a relatively small percentage of normal life among magpies. However, their variability, even if some events are rare, suggests something about the flexibility of magpie social relations and interactions with their environment and specific territorial conditions. Again, one may well be reminded of human social relationships.

Importantly, in the best defended territories many other bird species thrived in the presence of magpies and multiple species feeding and presence within such territories is common (Fig. 6.1).[29] Indeed, in the territory defended by the four adult males, the number of avian species dramatically increased within a 10-year period, with a noticeable increase in smaller parrot species and small songbirds, from pardalotes, thornbills, silvereyes, superb fairy-wrens, eastern spinebills, crested pigeons, to noisy miners and many others that made the magpies' territory their own and thrived under their guardianship. In such constellations, peace reigned and any inter-species conflict was near zero. Of course, such Arcadian idyllic situations can be very precarious and the bird assemblages can change in an

Fig. 6.1. Multispecies feeding (magpies looking up because of the photo being taken). This image shows magpies foraging together with galahs (two more hidden on the far right behind a shrub) and, barely visible, top right a crested pigeon. Just outside the image was a feeding grass parrot (red-rumped parrot) and in the shrub behind on the left upper part were two noisy miners, feeding on nectar-producing shrubs. All birds were relaxed and fed peacefully, even in rather close proximity to each other.

instant – it only needs storms or fire or building activities (such as road/infrastructure) and the balance is gone.[3,4]

It is important to remember that the faunal dynamics of habitat is partly determined by the faunal mix for the niche, no matter how complex that niche might be. Negative human intervention, introduced predators such as pet cats (in addition to feral cats), for instance, can have devastating effects on the sociality of peacefully cohabiting species, either eventually leading to an overload of predators or of the same cohabiting species suddenly turning on each other. We tend to blame the bird or even the species for 'aggressive' behaviour when, instead, we should be asking what has caused the problem in the first place (likely to be an outside source that triggered the change).

In well-balanced niches, most avian species live peacefully together. In those I have observed, some avian species even take joint actions. For instance, in serious cases of predator incursions, noisy miners joined the magpies in anti-predator defence. This happened in several cases when a goshawk, little eagle or wedge-tailed eagle had invaded the magpies' territory and the magpies were not able to stir, let alone move, the invader. With the noisy miners' support, even the most obstinate and determined predator chose to leave by the effective bluff of large numbers.

Magpies are largely sedentary birds and they seek to occupy a territory that will fulfil all of their needs for all seasons. Not all territories are the same and not all magpies manage to occupy an all-purpose territory and keep it. Finding a suitable territory with an adequate food supply and a suitable nesting habitat, however, is a primary condition for success in life, both in terms of the individual's own life span and health and in its ability to reproduce successfully.[5] The detailed investigations of magpie territories that we have (largely the work of Robert Carrick and of Jane Hughes and colleagues over the years),[6,7] seem to suggest, however, that the magpies holding such prime permanent territories may well be in the minority. Yet it is this group alone that tends to produce viable offspring.

Classifying magpies by territorial status

Robert Carrick introduced the concept of status according to territorial affiliation in magpies.[6] His subdivisions are still very useful today for understanding magpie behaviour in the broadest sense, their breeding dispersal and patterns, as well as their patterns of vocalisation and song repertoire. He distinguished between five groups.

First, there are the flocking birds that do not have territories but flock together in relatively large groups and feed in rather poor open spaces. They find a place to roost near each other but not in territories held by established groups. As flock birds, they generally do not breed; there are no fledglings or juveniles among them but often first year birds that have just recently left their natal territory.

Second, there are marginal groups whose conditions are worse than those of flock birds: they have spots for feeding that do not provide sufficient food and almost never places for roosting. Therefore, they need to traverse terrain occupied by other magpies either to roost or to seek additional food. They are barely tolerated and are attacked as intruders when they encroach on sites occupied by other magpies. These adults do not usually breed or, if they do, they fail. Marginal groups are smaller than those of flock birds and thus they are not able to muster enough support when under attack. One can see them huddling around trees or small sheltered areas (Fig. 6.2). Their life is extremely restricted, tense and unproductive and eventually, if their own survival is to be secured, they will have to leave.

Third, there are mobile groups that are forced to commute between poor feeding sites and small areas of roosting sites, and they tend to have to run the gauntlet between well-established groups. Any breeding efforts tend to fail.

Fourth, there are open groups and these are splinter groups from flock members who defend a feeding ground from others and may also prevent further flock members from joining, or evict present members from the group. They make no attempt to breed. Open groups are more frequent during the breeding season.[7] These groups are a little better off than the marginal groups, because they usually have more territory and their position tends to be more transitory, i.e. they will move on to better sites and these groups tend to form an interim stage to successful territorial occupation.

Finally, the only group that has all its needs met are those that have been able to establish permanent territories. Stress levels can also be high even in this group because any permanent territory is severely contested, at least for some time of the

Fig. 6.2. A marginal group. This group shows a behaviour typical of marginal groups: standing for hours 'hugging' and facing the tree, beaks often pointing at the bark or touching the tree and adopting crouching postures without feeding or drinking.

year, and its maintenance is a constant challenge. Some permanent groups are able to maintain their territory for a long period of time while others lose territories in less than five years. Nevertheless, it is only in this kind of group that offspring tend to be raised successfully to adulthood.[6,7]

Establishing a territory

Classic studies have been concerned with the biological significance of territories of birds.[8] The questions of territory size, shape, choice and neighbourhood have all featured in research and have been used to explain certain advantages and disadvantages of such arrangements. Even so, the processes by which territory is actually established and secured are still not well understood because so many variables can play a part.[9,10,11]

Establishing a territory can be very hard work, requiring continuous vigilance to defend and maintain it. There may well be a different set of problems in the centre of a territory as against the fringe of it (the centre-edge effect) and the territory may never be secure from being taken over by intruders.[12] Magpies who own permanent territories have the highest stakes and most to lose, and so will need to be vigilant all of the time.

One of the disadvantages of territoriality is the constant need for fighting or vigilance flying, activities that require constant high-energy output, perhaps higher than that required in a semi-nomadic or highly nomadic lifestyle. By opting to remain in one territory, magpies may have voluntarily forfeited some of the advantages of flight, such as the ability to move on and find better ground. The advantage of owning a territory, however, is security and predictability. The birds can breed and somehow seem to be able to adjust the size of their clutch in accordance with the quality of the territory, which is a considerable advantage. In the broadest sense, territoriality is a form of resource partitioning and may be the best way (for individuals as well as groups) of surviving in one locality and not another.

Magpies presumably establish a territory only when the site offers dependable resources, access to water, good nest sites and good roosting sites. How quickly and effectively magpies do this is not known and whatever other factors may play a role in territorial acquisition is also not well understood. One study suggested that magpies choose a site that is nutritionally advantageous[13] but it is not clear how magpies judge food quality and abundance and how they can accurately assess the benefits of one patch over another before settling. Most likely, the nearness of water, of open spaces with good ground cover and of mature trees are also factors in determining the best territory. The territory that has been selected may, in turn, affect group size or vice versa: group size may affect territory size. There is evidence that some groups succeed in eventually extending their territory to fit the group size and, equally, there is evidence that group size is determined by habitat quality.[14]

Once a territory is established, magpies settle for as long as they can. One territory in an outer suburb of eastern Melbourne, was held by the same male for 18 years. He was easily recognisable because of a slightly deformed foot causing him to limp and he, together with one (or more) female(s), raised brood after brood over many years. At a Canberra site studied by Robert Carrick, one pair was known to have held a territory for 11 years and the samples already mentioned had continuous occupancy by the same pair for the best part of two decades.[6,7]

Strangely, we have few studies of territorial birds showing how the decision of territory suitability is made and who makes it. Is a male investigating and then inviting a female in, are groups roaming the countryside to see what they can find or are pairs actively looking for a territory together? I am afraid we still do not have answers to these questions in magpies. In kookaburras we know that male and female together go on a search for suitable nest sites.[28] We know that flycatcher males have to establish a territory and the females, so it was found years ago, choose quality of territory over the male.[15] Komdeur[16] in the cooperatively breeding Seychelle warblers, *Acrocephalus sechellensis* showed some of the complications to consider in choosing a habitat. For instance, arising vacancies need to be of a certain quality because, as a rule, pre-breeding birds from territories of medium or low quality will only go to the same or higher quality.

Hence, if only low-quality territories are available it is better for the young bird from a high-quality territory to delay reproduction with greater lifetime fitness than those that disperse at one year of age and breed immediately in lower-quality territories. There has also been a wonderfully simple and elegant experiment conducted by Ward and Schlossberg[17] in Texas using the territorial calls of an endangered local songbird, the black-capped vireo, *Vireo atricapilla*. They showed that a simple territorial call played back in a vacant plot can attract pre-breeding birds to stay and establish a territory on the assumption that the presence of other birds of the same species would likely mean that the area is suitable.

Size of territory

The size of a magpie group's territory appears to largely depend on its ecological characteristics and the status of the group. In Robert Carrick's Canberra study the size of permanent territories varied between 5–60 acres (~2–24 ha) per pair or group.[6] Fertile grounds, good pasture and good roosting trees in a territory may require a smaller territory than poor open ground. Alternatively, small-sized territories can be a clear indication that the group occupying them may be transient, marginal and thus of low status. Marginal groups that fringe permanent territories may have to contend with less than do permanent residents – their territories may be as small as 2 acres (0.8 ha). In the Canberra study half of all groups were in this category but these were not regarded as permanent groups. Only 16% of permanent groups held territories of 5–12 acres (~2–5 ha), while the overwhelming majority commanded up to 30 acres (~12 ha).

Large sites of 12 ha of prime permanent territory require at least four to five magpies to defend them effectively and while smaller plots may be successfully defended for some time by one pair only, such low occupancy can only be maintained if there are no challenges from numerically stronger groups. The need for defence happens almost daily in the form of challenges presented mostly by intruding predators and by magpie neighbours. Indeed, the composition of the territorial group and its breeding success can vary substantially over time at the same site.

Territorial defence

As far as we know, the successful defence of a territory is dependent on many factors. A good territory may well be matched by good occupants. Plenty of experience, high levels of vigilance and strong group cohesion (effective teamwork) may be the most important qualities that mature individuals can bring to maintaining a territory. And group cohesion and teamwork may even be more important qualities than maturity and experience.

Defence of territory is related to two main sets of events imposed from the outside. One relates to magpie neighbours with either stable or marginal territory, not to magpie bachelor flocks that tend to change rather often in composition and by their very assembly is not stable.

The other is a response to intruders and predators. Intruders may also be magpies that are not neighbours. It appears that white-backed magpies intruding into the territories held by hybrid or black-backed magpies are dealt with more severely than black-backed intruders,[18] suggesting that plumage colour may play an additional role in territorial defence.

Neighbourhoods of strong permanent magpie residents are status equal and fights are relatively rare. Much of the territorial defence against magpie neighbours is conducted by one of the most energy-saving devices, that of vocalising. When a group is strong and has proved its mettle, it is usually sufficient, when having a neighbourly dispute, to voice a warning by a series of vocalisations referred to as carolling. One individual begins to emit the sounds and another one or two magpies join in chorus. The sound and structure of carolling is very distinct indeed (see Chapter 11). It has the sole function of proclaiming and confirming ownership of the territory and may also function as a reconfirmation of social bonds of the group.

Carolling has been identified as the magpie's most effective weapon in retaining territory (see also 'Carolling' and 'Duetting' in Chapter 11). In a study conducted by Robert Carrick, carolling was played back after removal of the dominant male and the territory was successfully defended simply by using the recorded vocalisation.[6]

Here, the 'dear enemy' concept might be applied.[19] While a neighbour may be viewed as a potential threat, a stable neighbourhood usually means that there are many battles that do not need to be fought. The groups have worked out their needs and the behaviour of neighbours tends to become predictable, particularly since individuals and their vocalisations are known, and some of those vocalisations are even shared.[20] However, the neighbourly dynamics change from an emphasis on 'dear' to an emphasis on 'enemy' when one group recruits more individuals to stay, either by reproduction or by actively allowing new individuals to join the group.

Numerical imbalance between neighbouring permanent groups generally leads to fights and to agonistic displays, often, so it appears, with the outcome of the larger group expanding its territory at the expense of the 'defeated' group. Some magpies from the larger group may set the stage for conflict by feeding in or flying into the edges of the others' territory and doing so repeatedly. Such behaviour is not tolerated and loud alarm calls made by the invaded magpies rally the troops together. The invaded group then flies to meet the invading magpies and, depending on the context, may employ a variety of strategies that are associated with several distinct displays. These displays are fascinatingly choreographed (see Fig. 6.3 below) and I have observed them in areas as different as Melbourne, Canberra, Armidale and near Coffs Harbour.

Tactic number one is the 'negotiating display'. The invading group looks on, usually from its side of the territory. The dominant magpies (one or usually two) assemble on the ground and walk along an imaginary territorial borderline (imaginary only to the human observer but obviously quite real to magpies) while the remainder of the group stays a little in the background. They pace up and down, like a border patrol. They fluff themselves up to look much bigger than they are and they throw their heads back and repeatedly vocalise in a manner that is solely associated with territorial defence carolling.[6] Several other magpies in the

Fig. 6.3. Boundary conflict and resolution. This is a rare sequence of photos, but quite a typical behaviour. A, Two magpie groups from adjacent territories approach and eye each other. The process took about five minutes. B, Here the two dominant males step forward and line themselves up in parallel to each other. Individuals named 'A' and 'B' are the leading males and both carol. The line emerging in the middle is a newly established territorial boundary. Individual 'Aa' belongs to the group on the left but, as an intimidation tactic, has 'crossed the line' and lines himself up with a magpie from the other group, also carolling, but 'Bb' does not reply (meaning that the will of the group on the left will be done). 'Aai', on the far left, is a relative of the group on the right but makes a firm decision to stay within the new group by walking away from its natal group. (C) Once the ritual is over, both groups begin to disperse and each walks away from the territorial border, without looking back at the other group.

invaded group may join in the carolling and such displays may settle the matter or be ignored by the invading group.

Tactic number two is the 'group strength display' – all magpies swoop down and then form a row at the border. This tactic appears to be used only when the invaded group is almost matched in numbers. If an invaded group is much smaller and has timid individuals in it, this tactic is not used.

Tactic number three is a 'leadership display' – the dominant male uses an aerial display to show his strength and agility, swooping up and down in the sky and then focusing his swooping at some of the enemy magpies.

Tactic number four is an aerial bluff – the entire group of magpies flying to the trouble spot doing aerial displays and vocalising continually. They certainly make their position clear. Usually peace is restored rather quickly but it may be an uneasy peace and sometimes the numerically weaker group eventually loses some ground to the neighbours. There is some indication that there are regional differences in the kinds of displays used. In Seymour, Victoria, close-range displays have been commonly observed, while in a Queensland site, aerial acrobatics are the norm.[18]

Once a conflict is resolved the territorial group assembles and, together, they tend to perform a long and loud vocal session of joint carolling. Here the carolling is not sequential but all magpies may overlap with one another making this a truly spectacular vocal display (Fig. 6.4).

Fig. 6.4. Group carolling. After expelling an intruder or having resolved a major conflict, adults congregate and carol vigorously as a sign of their victory and, presumably, their bond.

The function of any type of agonistic behaviour, it needs to be stressed, is not mindless 'aggression' but is regulatory for the very opposite purpose, namely to maintain harmony and collaboration within the group. Only if all members of a group fulfil their roles and can function as a team will the group be able to defend its territory successfully.

Exceptions and tolerated rule breaking

There are, however, many exceptions to the rule of vigilant defence of territories. For instance, one study found that magpies defend their territory more vigorously outside the breeding season than during the incubation and nestling stage of breeding.[21] Surprisingly, during the breeding season (when most swooping of humans occurs), birds intruding at the edge are often ignored altogether. Researchers explain that such oversight in territorial defence is likely to be related to the risk of egg or nestling loss. If a female has to leave the nest regularly to spend time defending the territory, eggs could be lost through loss of temperature or predator activity and the latter would also apply to nestlings.[21] However, this explanation cannot fully account for groups that have helpers. Presumably, in such social groupings, the brooding female can leave the territorial defence to others.

Territorial defence may be suspended altogether in exceptional circumstances. For several years in a row parts of the New England Tableland suffered a terrible drought and local creeks, the lifeline for stock and wildlife alike, began to dry up. The Kentucky Creek eventually consisted of just one waterhole, located in the central part of a territory belonging to one resident magpie family, which was known to have vigorously defended its territory over the previous five years. Two neighbouring groups now had no direct access to water. To my surprise, the magpies that had been so fiercely fighting for every inch of their territory over years suddenly stepped back and all the neighbouring groups were allowed to drink at the one remaining waterhole without the slightest conflict. It was a truce – the neighbours did not misuse the situation to gain more territory and the resident magpies made no single attempt to drive them away. All three residential magpie groups survived the drought years because of this ability to set aside normal rules applying at good times.

Since I was familiar with all three groups and had their wing-markings (as personal identifiers) on record, there was no mistaking that this truce that lasted over half a year was successfully upheld and enabled the survival of all until the rain finally arrived and made the creeks run again. I remain very impressed by this behaviour because it is so very rare. Indeed, I am aware of only one paper that focuses specifically on water competition in territorial groups.[22] In Wrangham's study of vervet monkeys almost 40% of the population of those without access to water died. It was not shared. Sharing a water source in land-based vertebrates seems to happen mostly when water is overabundant or the species sharing a water source are not territorial. Wrangham's study on territorial behaviour of neighbouring groups shows that under such tight competition for an essential resource, no dispensations were given among vervet monkeys.

By contrast, the suspension of the territorial rule by magpies to give access to water to all neighbouring groups seems an exceptional and perhaps unique behaviour that has never been described. It is possible that this may not be unique animal behaviour in Australia. After all, Australia has been drying out for 20 million years. Most animals and birds may have found ways of adapting to water shortages by giving up territorial behaviour for that reason (i.e. so they can continue to search for water far and wide). All animals must have water at some time. From my outback experience I know that when I see princess parrots or budgerigars a water source is a maximum of 10 km away, for zebra finches it is 2 km. And these may be locally privileged times and years when some waterholes do not dry out. For those very few species that uphold territorial rules all year round, as is the case in vervet monkey populations in Kenya[22] and in magpies in Australia, the matter of access to water becomes a matter of life and death instantly. At an interspecies level, it is extraordinary that these specific magpies chose to collaborate and not to compete – and all survived. The possibility that the magpies of the three groups may have been related to each other does not explain this behaviour because relatedness is not generally known to prevent fierce competition for essential resources.

Airspace rules

Magpies may also tolerate 'corridors' for safe passage through their territory. Two particular models are known to us. One is the passage via low flight to an outpost of territory that, by whatever mechanisms, the resident group has allowed marginal groups to use.[6] It is like a concession that permits fringe groups of magpies to feed in limited and designated areas of a resident group.

There are occasionally individual birds moving between territories – either because they are locally nomadic or are leaving their natal territory – that will fly at a considerably higher altitude than birds flying within a territory. It seems, although it awaits further study, that airspace may be subdivided vertically and higher altitude flight by other bird species (except birds of prey), well above tall tree lines, is generally tolerated by territorial magpies and leads to no vocalisations or active approach flights by the resident magpies. However, if altitude is lowered territorial magpies will respond immediately by aerial pursuit (Fig. 6.5).

Sometimes a strange bird or former neighbour may fly into a territory in order to stay. If this happens to suit the group (of more than two) the incoming magpie

Fig. 6.5. Aerial pursuit of a magpie stranger. Intruders are pursued on the ground and in the air, even juveniles will do so. Note that the juvenile (at bottom left) is about to attack the middle bird from behind. The other intruding juvenile (top right) has already achieved safe cruising altitude. (See also Plate 11, for aerial pursuit of a sea eagle.)

may not be attacked, be it a male or female. Acceptance or bare tolerance may be expressed in several ways. The best outcome is seen when the new adult may immediately proceed to feed and roost with the group. Within days, that individual will be incorporated into the group as if it had always been there. Rank decisions may be made in accordance with the size of the group and existing patterns of dominance. In the rare event of a dominant male being killed, under certain circumstances another male may assume his position.[23]

Usually, however, females or males will fit into existing structures and assume a subordinate role. The better the incoming magpie follows magpie 'etiquette' and does not challenge the superior rights of the dominant individuals (male or female), the less likely there will be agonistic encounters.

Defence against predators

It has taken some years, both of observation and careful experiments in the field, to arrive at the conclusion that anti-predator behaviour in magpies is highly sophisticated and based on substantial knowledge of the behaviour and skills of

Fig. 6.6. Local avian predators of birds that may take juvenile or even adult magpies. Top left: black-shouldered kite (*Elanus axillaris*); bottom left: goshawk (*Accipiter fasciatus*); centre: wedge-tailed eagle (*Aquila audax*); top right: little eagle (*Hieraaetus morphoides*); bottom right: peregrine falcon (*Falco peregrinus*).

predators that could become a danger to the adults and offspring alike (Fig. 6.6). We conducted studies over a period of five years in various territories from the New England Tableland (Armidale, NSW) to areas east of the Dividing Ranges. In all, 14 permanent resident family groups were tested using five different types of predator (avian and reptilian) known to be of varying degrees of risk to magpies and common in their habitat. In all, 210 trials were conducted using taxidermic models of the wedge-tailed eagle, a goshawk and a monitor lizard (*Varanus varius*). We found that magpies were very clever strategists who knew exactly how to approach and what to do to ensure that dangerous predators would choose to leave their territory.[24]

Black-shouldered kites include galahs and smaller parrots among their staple prey diet, which they catch during flight, but there is no incident known of them bringing down a magpie. Goshawks have been seen plucking live currawongs. Wedge-tailed eagles generally do not actively hunt magpies as they tend to prefer larger items such as rabbits but, occasionally, as I have witnessed, magpies do get caught and a magpie fits comfortably into one talon. Little eagles are avid bird hunters and so are peregrine falcons. These birds of prey not only differ substantially in size: the wedge-tailed eagle weighs 3–6 kg (males in the low, females in the high weight range), the goshawk is a slender, light-weight, lightning fast ambush hunter, the little eagle a wait-and-see predator and the solidly built peregrine falcon usually brings down its quarry at neck-breaking speed during flight. We were interested to see how magpies would deal with any such predators in their own territory. We were surprised to discover that magpies employed different techniques in mobbing each predator. There is a particular looping swoop they do when mobbing wedge-tailed eagles while they carefully avoid too close an encounter with goshawks, meaning they understand and correctly judge the abilities and strategies of their opponents[24] and skilfully remove themselves from any real danger. When the raptor has finally taken to the air, it will be accompanied by a magpie or two until it is well outside the territory (see Plate 11).

There is a further observation to be made in connection with anti-predator behaviour and this is a cognitive dimension. First, the skills and the coordination of mobbing efforts, executed as well as we found them to be, require skill and knowledge. The risk to the group has also fostered another extraordinary form of communication that has only been documented in the magpie and, later, also in the raven. This is the ability to communicate warnings about the presence of a predator by gestures understood by all magpies, as I discovered.[25]

Based on this discovery, I designed a number of field experiments involving the use of a taxidermic model of a wedge-tailed eagle. The wedge-tailed eagle was placed under a shrub so that it was not immediately visible and I then scored the behaviour of the magpies from the first moment of discovery by the first magpie to the moment when all magpies of that family were present. The first magpie, on discovering the presence of the eagle, not only raised the alarm using a specialised 'eagle' alarm call (more of this in Chapter 11) but, if the second magpie having arrived on the scene could not see what was wrong (clearly indicated by its scanning behaviour in several directions), the first magpie not only continued calling but resorted to a gesture of pointing. Without arms, however, pointing behaviour should have been impossible to perform in a bird. Instead, the bird used its entire body stretching it forward and down and using the outstretched neck, head and beak like a pointing arrow. The second magpie then followed suit and showed the same behaviour. Since it was not perched in the same position as the first magpie, its pointing towards the eagle had to be at a different angle and it was, i.e. it was not merely mindlessly copying the first magpie but chose angle and direction correctly to point at the eagle until such time as all magpies had seen the predator. This is an extraordinary behaviour.[25]

Gestures have an important place in speech development research in humans as a pre-speech but meaningful/symbolic act that is referenced for a particular declaration or outcome. Gestures were thought to be unique to humans but have now been studied extensively also in primates, specifically in great apes. And, of course, birds do not have arms or hands so a kind of referential gesturing was thought to be impossible. However, whole body movements and dedicated postures, as the work on Australian magpies[25] and common ravens[26] has shown, can fulfil the same function, i.e. such gestures are thought to require complex cognitive abilities in order to read their meaning: respond to the threat rather than looking at the other magpies.

Recently, a very interesting study by Mo *et al.* on mobbing behaviour of the powerful owl, *Ninox strenua*, has been conducted.[27] The powerful owl is a threatened species, the largest owl in Australia and one of the largest in the world with a slim distribution in NSW and Victoria. It is not only important to learn about the trials the owl faces in daily life, but it exposes very clearly that the magpie, often thought to be the only avian species, apart from the lapwing, *Vanellus miles*, engaging in mobbing behaviour is, in fact, just one of many native avian species using mobbing as a deterrent.

Their paper is also one of the few that defines mobbing behaviour using correct terminology, calling it an anti-predator strategy that contains agonistic responses to predators. Indeed, they observed about 30 species using similar strategies as the magpies, often not as sophisticated or as effective as those of magpies but in all cases the behaviour was due to the approach of a predator. By using the right terminology, it is made clear (as it should always be made clear) that mobbing has absolutely nothing to do with 'aggression' (more about this in Chapter 12). The mobbing birds are not 'angry', they are not 'nasty', they are not 'dangerous'. Instead, they have coevolved with these raptors and developed strategies to counteract potential attacks, avoid being eaten or having their brood destroyed. They want to move the predator away from their territory and also let the predator know that their location is known by all. Often this alone can spoil a hunt. The identification of the native avian species engaged in mobbing the powerful owl is also interesting because it involves very different avian families and birds of substantially different weight and size classes. There are large ones such as sulphur-crested cockatoos, ravens (three species), kookaburras, bowerbirds, noisy friarbirds, white-winged choughs, red wattlebirds and cuckoo shrikes but also species that are quite small such as many small honeyeaters, fantails and even thornbills.[27] I can confirm their observations because these species tally with my own species counts using mobbing against raptors. I might just add the yellow-breasted robin, like thornbills, a tiny but feisty and beautiful songbird that will also vocalise vehemently and even mob.

The mobbing behaviour likely to have the greatest success of all families of birds is that of the magpies. This is due to their weight class, acrobatic flying skills and their strategic approach to tackling any dangerous predator. Magpies are likely to be indispensable for the survival of smaller songbirds and parrots. It is almost superfluous to add that magpies are courageous, intelligent and persistent enough and, even more importantly, coordinate their efforts so extraordinarily well that they tend to succeed in driving a predator away. It is a very important point to stress that they achieve this by cooperation.

Cooperative species are defined as those species that help raise a brood – hence the definition of 'cooperative' is firmly tied to raising offspring. However, in the case of magpies, cooperation well exceeds brood care. The real success in the magpie's survival probably lies in cooperation in the field that involves every adult in the group (no juveniles are involved), males and females alike. Their utter reliance on each other, and their well-coordinated teamwork is the essence of their success in driving out predators often substantially larger than themselves. Similar examples can be found in wild dog packs and in lion prides. It is the precision of placement, of speed and of coordinating the attack in such a way that at least one member of the group can take over where the other left off. A comparison comes to mind with wolf packs and the high level of coordination they need to bring down an animal much larger than themselves. But here the comparison ends: wolves and other group-hunting dogs and cats are carnivorous and have learned these strategies to feed themselves. All the magpies need to achieve is to move the predator out of the territory. Many other species do mobbing flights against goshawks or whistling kites, such as masked lapwings, ravens, magpie larks, and even such small birds as willy wagtails and some honeyeaters will follow birds of prey in flight but these are for bluffing and warning and more often than not do not actually achieve the departure of the predator. In humans, the skills that magpies display (see Fig. 6.5) are usually the outcome of very specialised training in the military and in specialised police squads (I have your back – you have mine) which often takes years to perfect.

Fearlessness and teamwork are the secrets that enable magpies to rebuff predators of speed, skill or even weight – any of which can often be much more substantial than they individually possess. In over 90% of cases of mobbing behaviour in magpies, the mobbing birds do not actually make any direct contact with the bird of prey.

It is important to stress also that territorial behaviour and anti-predator defence are entirely independent of the level of hormones circulating in a magpie's body (gonadal state). That is to say, levels of testosterone do not determine the vigour of defence and defence is thus independent of the breeding season. In a test case, male magpies that were caponised (effectively the same as castrated) still retained their territory for years.[6] This is in contrast to nest defence. At that time, testosterone increases and the gonadal state, particularly that of the male, is linked to heightened alertness.

During breeding time, but also in large territories, adults may act as sentinels (Fig. 6.7) and warn others if a risk is spotted. However, any adult in the group can raise the alarm. Indeed, it is in the very essence of territoriality that those who lay claim to sole occupancy as a species and an individual group or family have developed a heightened sense of vigilance. A territory is hard earned and even more difficult to keep. Only those magpies that have fully learned what keeping it

Fig. 6.7. A sentinel choosing a vantage point. Note the stretched neck and erect posture, typical for vigilance behaviour. In exclusive pair occupancy, the male magpie will take over all guard duty, even when he feeds nestlings.

involves will reap its benefits, both in their own survival as well as in successfully raising future generations. It is a minority of magpies that achieves such status.

Endnotes

1. Durrant and Hughes 2005
2. Fulton 2006
3. Thorn *et al.* 2016
4. La Sorte *et al.* 2018
5. Dunn and Cockburn 1999
6. Carrick 1972
7. Carrick 1963
8. Hinde 1956
9. Dunham *et al.* 1995
10. Stamps and Krishnan 1999
11. Ahlering and Faaborg 2006
12. Johnson and Kermott 1990
13. Veltman and Hickson 1989
14. Hughes and Mather 1991
15. Alatalo *et al.* 1986
16. Komdeur 1992
17. Ward and Schlossberg 2004
18. Kallioinen *et al.* 1995 (Chapter 7)

[19] Temeles 1994
[20] Brown *et al.* 1988
[21] Brown *et al.* 1993
[22] Wrangham 1981
[23] Jones 2002
[24] Koboroff *et al.* 2013
[25] Kaplan 2011
[26] Pika and Bugnyar 2011
[27] Mo *et al.* 2016
[28] Legge 2004
[29] Bell 1980

7
Bonding and breeding

We do not readily appreciate how precious breeding birds are until we learn from the statistics that most never breed. We may not suspect how very special it is when birds succeed in raising offspring that are healthy and will breed themselves. While only a small minority of birds have surviving offspring, the lives of all other non-breeding or solo birds may hide records of tragedy, of partner loss or loss of offspring, droughts, starvation, accidents, unavailability of suitable nesting spots, disease and toxins. It is often assumed that any bird reaching sexual maturity will seek a partner and mate, raise offspring and do so year after year until some mishap or the reduction in reproductive ability stops that individual from reproducing. This is a myth. Evidence is plentiful that in some species over three-quarters of the individuals will not reproduce at all across a life span.[1]

In some species, including magpies, a breeding pair may constitute just 6–8% of the magpie population recorded in a specific area according to the *New Atlas* data.[2] Carrick suggested that only 14% of all magpies ever reproduce, a stunningly low number compared to mammals. This raises the question of why a few manage to successfully raise not just one brood, but a series of broods over a lifetime (in that they raise individuals that will also breed). Detailed research[3] has shown that it is not just a matter of the best and the fittest surviving and breeding, although a confident, experienced pair is obviously more likely to succeed than first time youngish birds. Environmental factors such as bad weather, accidents, unfavourable conditions, the effects of climate change and drastic reduction in number of nesting trees all play a part in lack of breeding success.[1,3] One can readily see, given these statistics, that maintenance of a species is not so easy and any prolonged problem can herald their decline. Yet, despite the evident precariousness of their lives and future as a species, birds have been highly successful in occupying every corner of the world. Magpies have managed to thrive in almost all areas of Australia, even in the hot Centre (see Plate 14). One of the notable quality about birds is that they tend to form bonds and provide extensive parental care to offspring. In altricial species, i.e. those that are naked and blind at hatching,[1,2] such as the Australian magpie, this includes the provision of a nest while the young are growing. Magpies give extensive care to their young, a care that does not end at their fledging but continues on for at least three months and parents provide protection and a safe learning environment for at least several months thereafter. Such commitment depends on both parents being present and committed to completing the rearing task.

Pair bonding

Birds frequently form long-lasting pair bonds.[4] This may not sound very remarkable but it is when compared to mammalian species. In mammals, only about 5% form lifelong pair bonds or even short-term pair bonds.[5] In birds, most species pair bond with a mate, if not for life, then at least in a bond that extends beyond courtship and copulation. Mate choice is thus a very large topic in avian ethology, yet absolutely nothing is known about how magpies choose each other.

We do not even know how magpie pairs typically meet. Magpies may 'meet' in flocks and then separate out as pairs in search of better pasture and a territory. Perhaps a lone female may venture safely into a territory as an immigrant provided it contains few or no females but several males. Alternatively, a male may displace another male and take over the resident female. Finally, female offspring may not leave the natal territory at all but become helpers for next year's brood and eventually breed within their natal territory having found a mate from another group. All examples and many other variants have been observed in magpies and are conceivable.

We also do not know if any element of choice is involved. For instance, do females choose males for signs of health such as feather condition, type of vocalisation and/or territory? Do males choose females or do females choose the males (as is the case in very dimorphic species – those in which male and female plumage and physical structure differs markedly)?

It is difficult to build a theory around mate choice in magpies as has been done for other bird species, because they do not have clear courtship practices and competitions (see more about this in 'Courtship' below). In other species, a male may produce special plumage effects, or he has developed special handicaps[6] (such as the long train of a peacock or the throat sack of a frigatebird), or has special breeding songs. This gives us some visible sign that females choose their mate according to selection principles.[7,8] In some species, ownership of a territory is more important to the female than the individual and some distinguish between territorial quality rather than the males, as is the case in female pied flycatchers (*Ficedula hypoleuca*).[9] Presumably, there are selection processes at work in magpies at an individual level but these may be subtle and they have been very difficult to observe. Extensive mate choice experiments have been conducted with other species and it appears that this is not straightforward because, in addition to all the possibly measurable criteria, there may be individual preferences. This is well known in the zebra finch. Zebra finches, *Taeniopygia guttata*, are socially monogamous and usually mate for life.[10] In this species, female mate preferences are predominantly individual-specific since several studies found little consensus in female choice. A recent interesting study showed that pairs that chose each other freely were far more successful in breeding than those pairs randomly chosen and forced together as pairs.[11]

We can speculate about several scenarios. Females may well have ways of assessing the overall health of a potential mate. Indications of health could be reflected in the plumage, in degrees of light refraction, in overall appearance, weight and confidence of gait. There is little sexual dimorphism in black-backed forms of magpies, however, so there is no guarantee that the choice may not be mutual, or that the male might also choose a female rather than vice versa. By comparison, white-backed magpies are highly dimorphic (males have white backs and females dark mottled backs) and here plumage patterns could play a role. Although there is no evidence one way or another, we do know, however, that once the choice has been made, female magpies need to extend a special invitation to the male to mate (see below).

What is clear, though, is that some magpie pairs may form long-lasting pair bonds and possibly even life-bonds and that some also appear to be monogamous, although evidence is mounting that high extra-pair mating rates occur regularly in magpies, in some places even at an astounding rate.[12] One of my own research sites in open forest hinterland near Coffs Harbour, New South Wales, is occupied by a magpie pair that has been together for 17 years – the entire length of our observation period. This pair has no magpie neighbours and their territory is well defined by forest edges. The nearest group of magpies is 5 km away and they too do not have neighbours abutting their territory.

Pair bonds in magpies, as already said, can vary as much as they do in humans. There have been reported cases of cooperative breeding[13,14] (see Chapter 9) and of both males and females being abandoned. Although there is no published evidence of magpie 'divorces', let alone the possible reason for them, my own anecdotal evidence suggests that they exist. The strategies for 'divorce' may vary also between magpies, as they do in other avian species.[5,15] In the Canberra and Snowy

Mountains region it has been observed that females who bonded with a dominant male often did not allow a subordinate female to breed. The breeding and dominant female intercepted the breeding process of a subordinate female at various times, either to actively destroy the other's nest during its construction, or her eggs, and even disciplined the breeding female herself,[16] making the pair bond an exclusive territorial prerogative.

Biological and social parenthood

Until recently, it was reasonable to assume that pair-bonded birds would both be the true biological parents of the brood they had raised. However, the newly available technique of DNA fingerprinting, which makes it possible to isolate the DNA and establish paternity reliably, has destroyed all notions of faithfulness and bonding as a precondition for social parenthood. DNA fingerprinting results have shown that even among bird species considered monogamous, such as the short-tailed shearwater, there is evidence of some extra-pair paternity.[17] Both male and female birds may have extra-pair coition producing fertile offspring for which the respective pre-established pair then cares.

Jane Hughes and her associates at Griffith University have been studying genetic relationships in magpies for the past decades. According to their findings, cooperative breeding of magpies is the norm, at least in Victoria (the Seymour area),[18] where magpies do not only engage in extra-pair matings but the female may actually lay her eggs in another female's nest.[19] This behaviour, more akin to that of a cuckoo female, raises many new questions. Does the female who so burdens another female with her eggs then mate again to lay another clutch herself? If so, this would be indeed a new twist to genetic assertiveness, so far presumed to be a prerogative of males in most species, with the exception of the well-studied superb fairy-wren. Here may be a species in which females too pursue a pro-active maximisation strategy of their own offspring by double-dipping and therefore the individual may raise more offspring with her genetic signature. In the Seymour area, extra-group paternity has been found to be high.[20]

Breeding season

Like most Australian birds, magpies have an extended breeding season. Depending on location, they may breed at any time between June and December, and in the Victorian and New South Wales Alps up to the end of January. In some areas, the breeding season varies due to climatic restrictions. Due to the rainy season in the north, subtropical and tropical magpies breed in the dry season (June to September), while magpies in temperate and colder climates tend to wait until August/September when the coldest spells are over.[21] Yet in terms of latitudinal differences of breeding times, there seem to be only small variations in peaks, onset and end. But even in climatically more restricted local areas there are at least three to four months in which a female can build her nest and rear her young, and such an extended breeding period can be advantageous. If, for any reason, a particular clutch has failed there may be a second chance to attempt another breeding cycle.

Introduced starlings have clearly learnt the value of an extended breeding season extremely well. In temperate regions they now arrive as early as June, well before any native species breed, build their nest and raise their young basically without challenge. By the time other species breed, the starlings are already on their second breeding cycle. One regular pair that I observed in 2002 achieved four rounds of breeding by December and eight surviving offspring. Magpies, by contrast, devote about eight months to raising their offspring from a single clutch, which means that they are slow breeders and will usually raise only one clutch per season. There are some exceptions, as the report on a plural breeding event (two nests in the same tree) by Western Australian magpies has shown.[22] At the other end of the scale, magpies and many other species may opt not to breed at all in a given year. This has been observed so far in extended droughts[23] but there may be other reasons as well. Some Australian avian species, such as the budgerigar, have even become opportunistic breeders. They do not have a breeding 'season' – their season is dictated by the rare rainfall of the inland but when it arrives, they will at once seize the opportunity.[24]

Courtship

I have not seen any records nor can I recall any encounters of seeing bonding female and male magpies that have involved visible courtship rituals or attempts by the male to impress the female with song, dance or other activities. The relationship of magpies seems to be rather devoid of all the drama that accompanies so many bird matings and partnerships. However, it is well known that territory itself can be the best trump card a male magpie can play if he is in the rare position of having secured an area without partner help. Even during the breeding season, the male tends to be busy defending the territory and, indeed, when the female is ready to mate (this is her call entirely), she has to place herself in a conspicuous position on a branch and call him over. Getting his attention can at times be difficult and she may have to make several attempts. She crouches on the branch, splays her wings outward and moves her fanned tail feather rapidly from side to side, in almost a figure eight type of motion. A complete cycle is about 1–2 seconds. The white tail feathers appear from under the black wing feathers resulting in flashing signals. Despite its signal strength, the male sometimes does not respond immediately and the female sometimes moves to a different location

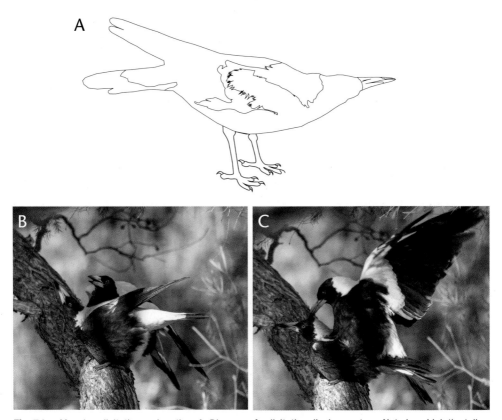

Fig. 7.1. Magpie solicitation and mating. A, Diagram of solicitation display posture. Note how high the tail feathers are raised. B,C, Rare photographs of magpies mating (source: Derek Midgley). B, The female is in solicitation display posture but, now that the male has landed beside her, she will position herself so that he can mate with her. She sits down. Note that she had chosen a wide-stemmed tree with rough bark. Sitting at a 90° angle to the branch, she has a firm grip on the bark, helping her not to slip or fall off when he mounts her. C, As songbird males have a cloaca rather than a penis, he has the difficult task of literally folding his abdomen down and around her tail to meet her cloaca. He stabilises himself by placing his beak on her head and using powerful wing movements so that he does not slip off. He will fold his tail end around her tail while she strains her cloaca to clear a space between feathers. When mating is accomplished, he immediately lets go and usually flies off. In this experienced pair, the entire process takes just seconds. Inexperienced males may well slip off several times.

and repeats the signal. This is referred to as a copulation soliciting display and has been observed also in female canaries (Fig. 7.1A).[25]

When the male magpie eventually responds to the female's call by flying in to mate with her, the encounter is brief (Fig. 7.1B,C). The only explicitly gentle courtship gesture that I have been able to observe is the male's bringing of food to the female and feeding her even before she starts incubating. This gift-bearing act by the male, common to many courtship rituals of long-bonding bird species (including kookaburras), is thought to cement the bond and apparently also signals to the female that he will be a good provider. However, in the examples that I have observed, such gift-giving happened *after* mating, not before. Therefore, it would seem that, in magpies, gift-giving is not an enticement for mating.

Nest building

Across all magpie subspecies, the magpie female seems to build the nest alone. There are no reports that a male has helped in actual construction of the nest. He may volunteer to bring some materials but he leaves the building to her. This suggests that magpie bonds are not true partnerships, as with long-term pair-bonding species, that share all tasks, including nest-building, and hints at polygamous arrangements. On the other hand, time devoted to nest building could be a distraction that is dangerous and sufficient to lose the territory to other watchful and waiting magpies. It is possible, therefore, that the division of labour between female and male magpies is merely an indication of the constant pressure on the pair to maintain a home territory.

Magpie nests are similar to those built by some birds of prey, and of similar size to those built by currawongs and ravens. The female places large twigs in semicircular fashion and piles on layer after layer until a relatively large but shallow cup shape is achieved. Nests of Australian corvids are almost indistinguishable from those of magpies or currawongs, although that of the Australian raven may be slightly larger. The defining characteristic of a magpie nest is the care that is taken in its creation: twigs are carefully arranged so that the strands can be woven together.

Nest construction materials include twigs, vines and larger sticks for the outer layer. Judging by the fact that vines and branches are chosen from a wider variety of plants than would seem necessary and picked from disparate locations, it is reasonable to assume that materials are carefully chosen and usually taken from the live plant rather than found on the ground.[26]

In areas close to human habitation, the outer layer of the magpie's nest may also incorporate wire, clothes hangers, fabric from hessian bags, binder (or baler) twine, silver paper, strips of clear plastic, rope, and even small adornments such as clothes pegs. The inner layer is like a second nest and much finer in structure. Materials used are softer and more densely packed. Here we find grasses, wool and other fibres, such as bark, moulded into a neat and compact circle with a slight downward sloping centre, so that the inner part of the nest is soft and well insulated against cold winds. Unfortunately, binder twine, nylon thread and fishing lines are also used in the inner part of the nest, sometimes leading to the offspring suffering horrific injuries.

In nesting sites that I have inspected, both in inland and coastal New South Wales, the nest usually incorporates the fork of the branch on which it rests. The centre of the nest can have a depth of up to 200 mm, in which case the female has usually filled up the fork with upward pointing branches. An upright branch helps to anchor the nest so that it is stable during any violent storms that may occur periodically at breeding time. This 'tying down' of the nest, necessitated by its exposed style (see 'Nest position and potential predators' below) is, indeed, an architectural feat. An inexperienced bird that may not have fully grasped the need for solid anchorage might therefore lose the nest.[26]

The inside of the nest is usually relatively deep, completely circular and beautifully lined with soft feathers and smaller twigs. At different sites that I inspected, the depth of some magpie nests led to additional occupancy by striated pardalotes, who dug a small tunnel into the lower part of the structure and lined it inside, thereby simultaneously achieving a safe and warm nest with roof. There was no attempt by the attending adult magpies to remove them and at no time during the breeding period did a magpie show any inclination to examine or intercept any of the pardalotes' activities within the lower tip of their nest. We do not know how common such pardalote occupancy is and in what sense it may be symbiotic. Clearly the pardalotes benefited having a safe haven and perhaps the magpies benefited by having a lower storey pest control built into their nest. Magpies are meticulously clean. Like many other bird species they remove their offspring's faeces, first produced in little sacks of their own and later the offspring are expected to raise their behinds over the edge of the nest and defecate well clear of the nest. But accidents do happen and their mere occupancy also encourages vermin to appear. Magpie females are rigorous and at least twice weekly will push their offspring unceremoniously aside and roughly go through the nest and clear out any debris.

Of course, all this sounds a little too much textbook and could suggest that all magpies behave in the same manner. It could also be (falsely) assumed that nest building is an activity that is genetically embedded and any magpie female is able to construct a nest. This is not so. Nest building is now often called 'physical cognition' because it is recognised that nest building is a difficult task that has to be learned.[27] Indeed, it is innovativeness in nest building that ranks highly in cognitive research on songbirds. In most species, such learning is either achieved by watching neighbours or even parents when youngsters stay on, or, as in so many cases, by trial and error.[28]

Inexperienced magpie females often do not at first succeed and may not succeed at all in the first season. They may not finish the nest before it disintegrates and then they may try again and even get as far as laying eggs – only to lose them through cracks in the nest or via a collapsing nest wall.

Fig. 7.2 shows another problem. Increasingly, in built-up areas native trees are becoming sparse or non-existent, having been replaced by decorative non-native palms. Magpies trying to nest in such positions, at least at first, can meet with failure. Unfortunately, fronds tend to dry out and drop off and those smaller ones

Fig. 7.2. First time attempt to breed. The entire branch broke off and the nest tumbled to the ground with it when the chicks were a week or so old. All three were killed as a result of the fall. It is unlikely that this magpie would ever choose a similar site again.

that support the fruiting bodies dry out even more quickly and will eventually drop off. This happens usually sooner than the nesting cycle can be completed, with disastrous result for the chicks.

There have also been quite a few reports of magpies building nests in unusual positions or making use of unconventional construction materials. There was one report of a nest being built in a flowerpot that hung under a breezeway in a suburban household. The more common experience is that people discover discarded nests that have a wire basis, constructed out of small soft wire, electric wiring and sometimes also involving binder twine. The latter has proved catastrophic for quite a few species of birds. If the binder twine is part of the internal lining the moving nestling may get the twine twisted around its leg and any attempt to free itself may cause the twine to cut more deeply, even amputating the leg eventually.

The worst case of binder twine I have witnessed was the case of a magpie nestling who, when trying to fledge with the others, was thrown out of the nest and caught by one leg and then hung upside down dangling helplessly from the twine, head down from the nest. The distress calls alerted people who were kind enough to ring for help. The electricity company also owned the only cherry picker in town and they were willing to come and help rescue the bird. It was the only time I have had a joy ride in the basket of a cherry picker. I was armed with just one syringe of saline solution to rehydrate the bird and a sharp, fine pair of nail scissors and antiseptic drops. The cherry picker managed to get close enough to the nest that I could rescue the bird from its life-threatening upside down position, investigate the damage and manage to remove the binder twine, all while swinging mid-air in this basket high above the road. Although the flesh was cut deeply, the bird was able to move the foot. Then the bird was rehydrated and placed back in the nest after I had removed all traces of the twine. This task was barely completed when the rehydration took effect and the frightened and exhausted bird suddenly sprang to life, fluttered and flew straight to the waiting mother and even landed successfully on her branch.

It was an occasion that had many community members involved and filled us all with joy. People even clapped. If one ever asked me how humans should interact with native wildlife I think I would be tempted to choose this example. By the way, neither parent tried to intervene, let alone swoop at me, although such a move would have seemed perfectly acceptable to me under the circumstances.

Nest position and potential predators

One astonishing element in magpie nest building is their positioning, which usually gives them full exposure to the skies and the sun. Of the hundreds of magpie nests that I have seen (in Victoria and New South Wales) only some have been protected by good foliage. The majority have been found in the most exposed parts of a tree or at least in parts with a clear opening to the sky. While this may prevent surprise visits from goshawks, an ambush species, it leaves the nest and the nestlings open to other avian predators and the elements. Ravens, for example, will readily take magpie young when a nest is unattended.

Nest height is partly determined by the height of the vegetation on offer. In more arid regions, magpies may have to contend with fairly low branches of gnarled shrubs and trees, almost without leaves, no more than a few metres above ground and almost within human reach. In a rare case reported from the Little Desert National Park in western Victoria, a magpie nest was found less than a metre off the ground, despite the presence of trees.[29] My research on vocal development in magpies[30,31] required that I was at least at eye-height of the nest and in almost all cases that could be achieved by standing on a 4-m guy-roped ladder, hence, with my own height added, about 4–6 m off the ground. In all these cases, nest sites were in deciduous trees or pines. In mature gum trees, nest height tends to increase substantially because the smaller forks and branches suitable for anchoring a nest are only found further up in the tree, at a height of ~10–15 m. There does not appear to be an ideal height for nest building. Considerations seem to centre on the need for anchorage of the nest. However, I am not so sure whether experienced birds do not watch for wind direction. I have found that magpies tend

Fig. 7.3. Successful nest site choices over 10 years for one pair. The magpies chose different nest sites in the same group of pine trees but in slightly different locations: left (used two seasons), middle (used for five seasons), right (two seasons). In the first year the female chose a nest site in a slender tree (third from left) and this nest was destroyed (hence not marked) and no further nesting attempts occurred that first year.

to build downwind and, therefore, are more likely to gain maximum protection from the tree in which they have built the nest, the tree itself providing the shield. This was certainly the case in one magpie territory that had three full-grown pine trees. The female reused one nest site for several years but over a 10-year period changed nest positions twice, each time on the downwind side facing east in the same group of trees that had seen her raise so many offspring successfully (Fig. 7.3).

In several parts of Australia one of the magpie's main predators is a varanid (monitor lizard). In areas where monitor lizards are numerous, magpies tend to shift nests to outer branches. In a fine balancing act of survival, magpies may breed during the colder part of the season when monitor lizards and snakes are either still hibernating or only emerging for brief periods of the day, thereby risking the survival of the clutch due to inclement weather. However, the female is very protective of the nestlings in the first weeks and will shield them from too much sun and rain, spreading her wings right over them during downpours and when nestlings start panting due to too much exposure to the sun.

Fig. 7.4. Rare photograph of a pair together in the nest, female on the left (note the grey feathers on the nape of her neck), male on the right (clear white feathers at nape of neck). It is a gesture of affection and probably an encouragement for her to remain on the nest since she does all the brooding. Although the quality of the photograph is not high, it is included here to illustrate the male's presence in the nest.

Of potential nocturnal predators, it is known that the barking owl will prey on both magpie adults and offspring. It is difficult to think of a context in which the magpie's spectacular and conspicuous plumage could offer camouflage by day. It may be useful at night, however. The dappled light in trees hides a black bird well and perhaps particularly well if the shape is broken up by white patches that then may appear as pools of light. Such camouflage would work better in the open than in completely light protected areas. However, this is mere speculation and would depend on how threatening nocturnal predators are, or maybe once were, for magpies. Today, the barking owl, like many birds of prey, is a threatened species and encounters between magpies and barking owls may be rare.

Nest building is clearly shaped by evolutionary and ecological pressures. At no time is a species so vulnerable as when it is trying to raise offspring. We can therefore assume that the nest building strategies that magpies have adopted are somehow beneficial and will maximise the chances of successful breeding for the individual.

Brooding, clutch and clutch size

When the female is ready to lay eggs and shortly thereafter, the male may even join her in the nest and sometimes even groom her, acting very gently near and next to her. As far as we know the male does not undertake any brooding even when he sits next to her (Fig. 7.4).

While the female is brooding she may rarely get up, let alone collect food for herself and the male gets a firsthand experience in feeding because he now has to feed his female partner on the nest.

The clutch size of the magpie is usually listed in bird guides as consisting of between two to five eggs; one guide has made the extraordinary claim of six.[32,33] It takes about one adult magpie to raise one offspring, two adults for two and so forth, so clutch sizes larger than two may require the support of helpers. Of course, additional eggs may be laid for insurance purposes in some species because not all eggs may hatch, but we have no evidence that magpies adopt this kind of breeding practice.

In Victoria, I have observed magpie groups raising as many as four offspring within the one territory. In New South Wales the number of hatching eggs seems to converge rather towards two with a maximum of three. In cooperative arrangements, as many as four magpies may be raised in one clutch but I have never come across an example of five or six offspring being raised successfully in one clutch. In Robert Carrick's study in the 1960s in the Canberra region, the overall mean clutch size was 3.5.[34]

Clutch size depends on many factors: seasonal variations, weather conditions, terrain, status, health of the breeding female, available support to help with raising the young, food availability, predator presence and even effects of parasite overload. For example, hot dry summers lead magpies to abandon nests. Large clutches of more than three may lead to parasite increases which, in turn, affect food consumption and energy demands, leading to poorer health quality of the offspring.[35] Research has found that the female of most bird species is able to make decisions about the clutch size[36,37] and it would seem reasonable to assume that the same mechanisms may also be at work in magpies. There are many reasons why a clutch may fail. Magpies have natural predators and nest parasites, such as the channel-billed cuckoo, *Scythrops novaehollandiae*, the largest cuckoo in the world. It prefers to lay its eggs in magpie and currawong nests. Magpie parents will protect the channel-billed cuckoo chick and feed it but, as a result, they will not be able to raise their own young. The channel-billed cuckoo's brood parasitism is now less apparent because the bird is no longer as common as it once was in its eastern inland distribution. However, the common koel preferably parasitises magpie, currawong and red wattlebird nests and the hatched cuckoo will expel all other nestlings. Such risks are ever present.

Survival rate

The goal of any individual and ultimately of the species is to survive. Not surprisingly, survival rates are a measurable unit to establish who can and will raise offspring that will themselves also be successful in raising offspring. Such queries have made mate choice a crucial topic that has been around for a very long time. Of late, however, the old incompatibility and compatibility hypotheses have given way to a slightly more open approach suggesting that some birds can actually like each other and are compatible for very individual reasons. As one recent study of zebra finch partnering behaviour found, coupling up pairs randomly (and without the birds' consent and wish) had poor results. In females, we observed a reduced readiness to copulate with the assigned partner, while males that were force-paired showed reduced parental care and increased activity in courting extra-pair females,[11] all situations that have a direct effect on the quality of parenting and the time devoted to the offspring. Therefore, it is not too far-fetched to assume that well-matched magpie couples may stay together longer, perform better in their territorial defence and also raise more youngsters more successfully.

A set of studies recently addressed the question again as to the importance of latitudinal gradients and the overall survival rate and life history of birds. It is now well known that Australian birds of any weight class tend to live a good deal longer than their Northern Hemisphere counterparts in similar weight ranges. Of course, Australia's share of the tropics is relatively small and magpies occupy vast areas of temperate climate (see Plate 2). We know that, generally, organisms living in hot climates and are sedentary have a lower metabolism than those in cooler climates. Lower metabolism generally means a chance of a longer life span than a high metabolic turnover. A European study tested this assumption on a small songbird species, the European stonechat, *Saxicola torquata*, which happens to have a population in Ireland that is half-sedentary and half-migratory (overwintering in Spain), two migratory populations, one in central Europe (overwintering in North Africa) and one in Kazakhstan (overwintering in Northern India) and one sedentary population in Kenya and they found the gradient principle of metabolic rates versus latitude confirmed: the Kenyan population of stonechats had the lowest metabolism, the Irish sedentary had a mid-value and the two migratory groups of stonechats from high latitude areas had the highest metabolism of the four groups of stonechat, even though all four test groups were held captive in the same environment.[38]

Another long-held assumption was that clutch sizes get larger, the higher the latitude but the number of broods per year gets higher the lower the latitude. In other words, in birds living in lower latitudes, as most Australian birds do, compared to northern American and European latitudes, each clutch should be relatively small but then could be followed up with another clutch in the same season. This is not altogether incorrect. One group of studies choosing bird species from north and south of the equator in the Americas found some confirmation that temperate birds have lower survival probabilities than their tropical counterparts.[39]

How does this help us with understanding clutch-sizes and survival of juveniles and even adult longevity in magpies? The answer is probably relatively little at this point in time. Magpies have a long life span reaching probably 25–30 years overall (suggesting a low metabolic rate).[40] The study by Gibbs[21] was important in that it provided quantitative evidence of widespread and systematic differences in magpie breeding patterns over geographical and temporal climatic gradients. It also established that timing and magnitude of magpie breeding varied annually in response to climatic conditions but, to my knowledge, we have no data on metabolic rates across the same gradients. It would be quite important to study survival in pre-reproductive birds as Tarwater et al.[41] did, choosing a neotropical passerine, the western slaty antshrike, *Thamnophilus atrinucha*, in central Panama. They report that fledged antshrikes had 76% survival through the dependent period and 48% survival to the age of one year; survival rate was lowest during the first week after leaving the nest. Timing of fledging within the breeding season, fledgling mass, and age at dispersal influenced survival, while sex of offspring and year did not. Individuals in this Central American species did not breed until two years of age, and post-fledging pre-reproductive survival was 41% of annual adult survival.[41] Magpies usually do not breed until five years old. Such studies would be useful for the Australian magpie since the overall theories on predicted survival rates in high and low latitudes developed in the Northern Hemisphere do not provide a reassuring fit for this widespread bird across a continent that is largely not forested but dry, not predictable but fickle. It could help us understand better how parenting but also latitude and climate generally affect survival rates in magpies.

What we do have, even though relatively localised, are the studies of Robert Carrick and his colleagues in the Canberra region.[34,42] They provide the most useful benchmarks for estimating the survival rates of magpie offspring. In their research, they counted all surviving magpie offspring in a given season and also took note of territorial quality. From this, they produced evidence that, in good permanent territories, the survival rate was only one magpie offspring for every two adult hens; or, put another way, for every seven offspring hatched in that season there was only one that survived to adulthood. These results suggest that there is an extremely high attrition rate of 86% of offspring of all breeding magpies (not counting any non-breeding populations) or a survival rate of just 14% for all broods of magpies occupying good territories. If we apply these calculations to an active breeding span of, say, a maximum of 20 years for one adult hen (most likely to be less), with a total of 60 offspring hatched (assuming three per annum for 20 years) under extremely optimal conditions, then an adult female may produce an absolute maximum of 8.4 surviving offspring with a chance to breed themselves.[34,42]

For those magpie groups in peripheral territories, of poorer quality and often offering less shelter, the reproductive rate, according to the Carrick studies, declines to just one surviving offspring per 11 females; or, of 39 young magpies hatched that season only one survived to breeding age. If, for argument's sake, we assume the same life span in these females as in well-fed and well-positioned magpies, the reproductive rate would not quite reach the number of two viable offspring per female over her entire lifetime.

Yet, even with these calculations, we do not know what the overall reproductive rate of magpies is. The high attrition rate of nestlings and juveniles is measured in a context of a pool of select magpies that are able to breed at all. However, the slow rate of replenishing the magpie population seems to suggest that overall population viability depends on the longevity of each individual that has reached breeding age. Hence, it is probably a good deal more difficult than we have surmised for magpies to raise and keep alive an upcoming next generation.

Endnotes

1. Sternberg 1989
2. Barrett *et al.* 2003
3. Newton 1989
4. Black 1996
5. Milius 1998
6. Zahavi and Zahavi 1997
7. Armstrong 1965
8. Bennet *et al.* 1997
9. Alatalo *et al.* 1986
10. Riebel 2009
11. Ihle *et al.* 2015
12. Durrant and Hughes 2005
13. Cockburn 1996
14. Schmidt *et al.* 1991
15. McNamara and Forslund 1996
16. Frith 1969
17. Austin and Parkin 1996
18. Finn and Hughes 2001
19. Drayson 2002
20. Hughes *et al.* 2003
21. Gibbs 2007
22. Fulton 2006
23. Veerman 2003, pers. communication
24. Pohl-Apel *et al.* 1982
25. Leboucher *et al.* 1994
26. Baldwin 1979
27. Auersperg *et al.* 2017
28. Guillette *et al.* 2016
29. Gardner and Gardner 1975
30. Kaplan 2018a
31. Kaplan 2018c
32. Simpson *et al.* 1993
33. Reader's Digest 2002
34. Carrick 1972
35. Poiani 1993
36. Lundberg 1985
37. Martin 1992
38. Wikelski *et al.* 2003
39. Muñoz *et al.* 2018
40. QNPWS 1993
41. Tarwater *et al.* 2011
42. Carrick 1963

8
Caring for the young

The magpie breeding season is now so much part of Australian life that it has almost become a public debate. It is discussed on radio, as well as in newspapers and receives countless comments on social media punctually every year around September when some male magpies are seen to actively defend their nest site. There is no single other species about which people have expressed such strong and public interest and where tempers can run so high, both positive and negative. In 2017 the magpie was declared the most popular bird in Australia by a survey conducted by *The Guardian* and BirdLife Australia, reflecting the status of the Australian magpie as a national icon.[1]

For the unsuspecting magpie at least, late winter and spring months are very important because it is the time of nesting. Those magpies that have made it to this stage may well have a reason to celebrate. Many things can still go wrong in the following months, but the right preparations have been taken. The commitment of the pair is in place, the nest built, with some luck, food is relatively easy to find and the male will now step into a new role to be played out in three acts.

A good male will regularly feed the female while she is brooding, reassure her by stopping by at the nest and even sharing the nest for brief periods (as was shown in Chapter 7). While demonstrating that he is doing his job as a sentinel and has assured her by feeding her, eventually the brooding will end and the young will hatch. Then he will have a new job during the time the young are confined to the nest and that is to spot any potential problems or threats. Even if other younger magpies share the territory with him, he is basically on his own and has to demonstrate to the female that he is doing his job of providing safety and food for the female and the offspring, running solo at least for two weeks before the female might consider short spells off the nest. And when the young finally fledge, he is relieved of his defender role. At this time both parents will frantically provide food from dawn to dusk for months to come, introduce the offspring to the territory, lead by example and guide their young safely through the many risks and dangers that they might face. Females seem to have rather strong opinions because if he sits near the nest while there is perceived trouble she may vocalise in short commands until he does something about it. This has been observed several times. It seems that the female has considerable 'say' in his behaviour at this time.

The active physical and social development of magpie chicks requires an investment by the adults of at least eight months. Magpie fledglings are fed for three months by parents and, where present, helpers, then closely accompanied for another three months, after which the juveniles are supervised for yet another few months. If offspring stay on, some of them may become helpers for next year's brood and thus watch and observe how nests are built and young are being raised.

Although magpie females are biologically capable of reproducing (sexually mature) by their second year, there is usually a substantial gap in time between dispersal and first breeding attempts – often three to five years or even longer. The same is true of males. Ultimately, it is not biological reproductive ability but knowledge, skill, confidence and vigilance (and some luck) that will determine whether a pair is ready and able to reproduce successfully and raise young.

Egg and incubation

Magpies are an altricial species – which means that embryos hatch at a relatively early stage of development (eyes are closed at hatching, the body is naked and the small body barely has the strength to raise its head).

Hence, the chicks are completely dependent on parental care and protection. Magpie eggs are not very large but can be as much as 27 × 38 mm (Fig. 8.1). The eggs can have different shades of colour from light blue or green to blotchy, some even have reddish colouring.

Eggs size (as well as weight) depends partly on the size of the female who laid it. Females in better condition lay larger eggs[3] and these are more likely to hatch than smaller ones. When a graph is plotted of the average egg weight in various species belonging to the same order against the average weight of the females of those same species, a clear mathematical relationship is found between increasing egg weight with increasing bodyweight.[4] Put simply, large females lay large eggs and small females lay small eggs.[2] One would expect eggs of *Gymnorhina tibicen tyrannica*, the largest magpies in the south-eastern corner of mainland Australia, to be slightly larger than those of the relatively smaller Northern Territory subspecies, *G. t. eylandtensis*.

Fig. 8.1. Magpie eggs. It is not uncommon for one or two eggs to be infertile. By contrast, the chicks of precocial species are able to walk and feed themselves almost immediately after hatching. The eggs of altricial species are smaller than those of precocial species – they are relatively lean with sparse energy supplies.[2] For instance, the egg of the precocial masked lapwing, *Vanellus miles*, a bird of similar size to the magpie, is 43 × 49 mm. Incubation in lapwings is a week longer than in magpies but the emerging hatchling can feed itself within three days. The altricial magpie can usually feed itself only after three months or even later.

Egg weight to some extent determines the survival of the hatchlings,[5] as well as their long-term survival and reproduction. As a general rule, the heavier the egg and the more nutrient reserves are available, the heavier the chick will be at hatching and the more likely it will be to survive to adulthood.[6] Moreover, not just the female but the preferred males may, indirectly, increase egg size and weight and the chances of the offspring surviving, because males in larger or better territories are better able to provide the female with high-quality nutrition during nesting. In other words, the flow-on effect of a healthy and well-fed parent pair is enormous even before the chicks hatch. Ultimately this is related to the quality of the territory itself.

Incubation usually takes 20 days. Females brood the eggs continuously and do this without male help. In well-bonded couples the male will feed the female, particularly during the first week of incubation. If weather conditions are right, she may briefly leave the nest to seek her own food but, in all instances that I have observed, she will swiftly return to the nest. Most females do not leave the nest at all, just occasionally raise their bodies and rearrange themselves on the nest. While she leaves the nest, the male or one of the helpers or members of the territory (even if not a helper) will stand guard and, as a sentinel, seek a high vantage point.

The development of the embryo inside the egg follows a pattern similar to other altricial species. The embryo obtains from the egg yolk the nutrients that it needs for growing and forming its various specialised tissues. The female even deposits a certain amount of her hormones in the egg and these can affect the development of the embryo and the behaviour of the young after hatching. As the embryo develops it starts to move inside the egg, reaching maximum mobility about halfway through the incubation period.[7] Wings, legs, head and beak all move and sometimes the whole body turns inside the shell. This activity is essential for the developing nerves and muscles. In the last few days before hatching the embryo will react to external stimuli, such as sound, touch and light,[8] and these may have an effect on the magpie embryo.

Hatching

Hatching requires a large amount of energy and coordinated movement by the embryo. It may take hours for it to make a small break in the shell using its egg tooth, a sharp point at the tip of the upper mandible of the beak. When the end of the egg has been cracked, the young bird completes the hatching process by pushing with its feet and body until it is freed from the eggshell. Altricial hatchlings cannot regulate their own temperature at first and are utterly dependent on the warmth and the nutrients that the adults provide. When they get cold they either crawl under the hen or they make special calls that elicit brooding by her.

In magpies, hatching of each egg in the clutch occurs at about the same time (synchronous hatching). In predatory species (eagles and others) as well as in tawny frogmouths[9] and kookaburras, each egg in the clutch hatches at a different time (asynchronous hatching), and can take as long as three or four days from first to last hatching. If hatching is synchronous, the offspring are at the same stage of development and they continue to develop simultaneously. In asynchronous hatching, the slight developmental difference among the offspring,

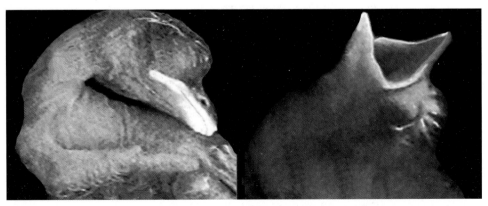

Fig. 8.2. Magpie hatchling, the first 24 hours. Left: shortly after hatching the tiny hatchling is exhausted and sleeps in the foetal position. Within 24 hours it can very briefly raise its head to receive its first meal. Note the eyes are firmly closed.

tawny frogmouths exempted,[10] can lead to conflict and competition between older and younger siblings and even to siblicide when food is in short supply. Such sibling competition does not occur in magpies, although it will invariably occur when people hand-raising magpie nestlings put individuals of slightly different age together.

Magpie hatchlings are at first naked, pink-skinned and blind (Fig. 8.2). They have inordinately large feet, a very long neck and a relatively short but broad beak ending in flexible yellow cartilage enabling the bird to open its beak very widely, called a gape. The inside of the beak is bright red. Magpie hatchlings open their eyes fully ~10 days after hatching and they begin to show fine downy feathers on head, wings and back in the first week. By the second week pinfeathers develop on their wings and tail, and there are downy feathers mixed with pinfeathers on the rump while their back is still entirely without feathers. The feathers on the head are already black and cover most of the head. Even at the very early stages of development, the typical black and white plumage begins to show.

Parental care after hatching

The magpie is among the few birds that feed their offspring equally and consecutively; others include the budgerigar (*Melopsittacus undulatus*), white-winged choughs (*Corcorax melanorhamphos*), tree swallows (*Tachycineta bicolor*), pied flycatchers (*Ficedula hypoleuca*) and crimson rosellas (*Platycercus elegans*). In many other species, parent birds respond preferentially to the most effective begging, causing a feeding hierarchy that favours the stronger nestlings.[9]

Magpies feed their nestlings a rich and varied diet and they do so at a rate at which nutrients can be harvested, usually every 20 minutes. Among the food items offered, earthworms tend to appear frequently on the menu (Fig. 8.3). Earthworms have very little nutritional value but they store substantial amounts of fluid and may be important for this reason alone. Magpies do not start drinking water until they have fledged and my own observations would suggest that such drinking does not commence until several weeks after fledging.

Usually, on arrival at the nest, a magpie parent has only enough food for one nestling and either needs to return to collect more feed for the other nestlings or a partner or helper feeds them. Magpie nestlings stretch the neck, gape the beak and make faint begging calls. Immediately after hatching, when their eyes are still shut, the begging response is triggered when the landing parent causes the nest to move abruptly. This vocal begging behaviour increases in length and loudness over the weeks. In the first weeks of life, the begging calls are usually brief and high-pitched faint peeps, by the third week they have become louder and begin to sound more characteristic of magpie vocalisations. By the fourth week calls are audible over long distances and the nestlings are alert and curious about everything (Fig. 8.4).

Fig. 8.3. Dinner is served. The male has collected two items to feed to two nestlings. In this particular case, the adult can actually separate the two items and feed the earthworm first (with some assistance from the nestling) while holding on to the cricket which can then be fed to the other nestling (Photo: Kim Wormald, www.lirralirra.com).

Later, when their eyes are open, the same begging response is triggered by the arrival of the parent, either perched on the edge of the nest or at a branch near the nest. By the third week, nestlings scan the environment to monitor the whereabouts of their parents (and their helpers) and begin to adopt a feeding posture well before an adult arrives at the nest. Nestlings will stop begging once feeding begins or when the food disappears in the beak of a sibling. It is only in the last two weeks that one can comfortably hear the begging calls of the youngsters, because their voices are now strong enough to carry over longer distances.[10] The change of strength of the voice has to do with the extraordinarily rapid growth they undergo in the last 10 days before fledging (see Table 8.1). In that same period, the nestlings begin to take a great interest in the world around the nest and watch keenly where their parents might be.

Although all nestlings vocalise briefly when a parent bird reaches the nest's edge, the adults appear to make the decision on which one to feed. This decision seems not to be made on the basis of begging calls or physical shuffling of the young for the best position, but on the basis of who has not been fed before. The patterns of feeding behaviour, of course, determine the rate of growth and so all nestlings grow together at the same rate.

Feeding habits of offspring, as described in Chapter 5, also seems to be under some regime of selectivity in the sense that the parents seem to understand something about nutritional value or at least about food that is more filling/ worthwhile than other food. In an experiment conducted by Larkins[11] adult parent

Fig. 8.4. By the third week and as nestlings grow they also start to take an increasingly active interest in the environment outside the nest.

magpies were presented with morsels of different foods. He found, for instance, that the adults did not feed cheese to nestlings but chose mincemeat for them instead. Adults were leaving insects behind but feeding scarab larvae to the offspring and only keeping the remainder for themselves. This does suggest some ordering principle. It could be the weight, the taste or the size – whatever it was in the small set of observations that Larkins made so many years ago, it opens a window into the way magpies may think before they proceed with feeding. This is certainly a behaviour that would merit further (systematic) study. By whatever criteria the birds singled out the food for their offspring, the choices they made raise several questions about what they really knew about the food they ate, precisely because they were so consistent. Parent magpies could easily judge correct amounts to give because satiated nestlings will simply not beg for food. Presumably harmful food may be known and can be judged and avoided by colour, taste or smell but anything edible would also bring in the question of quality of food.

When I hand-raised magpies, I compared the weight, beak length and other measures with nestlings of wild-raised magpies and as pointed out elsewhere,[12] the hand-raised magpies had consistently shorter beaks than the wild-raised ones. It is conceivable that magpies feed items in their diet to offspring that even the most carefully designed diet administered by humans cannot reproduce and the human-designed food may well be lacking in some vital ingredients.

The 'fairness' of magpie parents towards their offspring can also take extreme forms. At the time of fledging, all youngsters tend to leave the nest together. At one nest site I observed, the second nestling had failed to leave the nest when the first one had fledged. The parents continued to supply food to the nestling first and then, quite a distance away, to the fledged offspring. They maintained this feeding schedule for a week. I then intervened and found a piece of plastic twine had tightly anchored one leg to some embedded twigs within the nest so that the young magpie had been unable to move from the nest. After freeing the bird from its shackles, it was at once keen to leave the nest. Both fledglings continued to be very well fed, showing the fair parental treatment of their offspring, something not necessarily expressed in other species.

Housekeeping and protection of nestlings

The younger the nestlings are, the longer are the periods the mother stays with them for reassurance, warmth, and protection from sun, rain, wind and from predators. Also nest hygiene matters to magpies. Indeed, nest cleanliness is a high priority. In the early stages, the nestlings' faecal waste comes inside a gelatinous sac which the parents remove immediately – the magpie parent is ready to collect the faecal sac as soon as the nestling expels it and before it even touches the nest. The very young nestlings, before opening their eyes, do not defecate until they feel the nest vibrate (which indicates that a parent has arrived). Later on, the magpie parent may actually prompt defecation by tapping its beak directly at the offspring's anal region.[13]

The female magpie maintains nest hygiene and tidiness in a daily routine that involves rearranging the nesting material, removing any debris and checking for any protruding sticks. This activity is quite vigorous, involving pulling, pressing and throwing unwanted debris over the nest's edge. While this cleaning activity goes on she directs the nestlings to one side of the nest and then moves them to get to the other side.

Nestlings are very vulnerable but they are also constantly hungry. This, in a way, places contradictory demands on the parents – they have to leave the nest to collect food but they also have to be at the nest site to protect the young and keep them warm. In the first two weeks after hatching, the male (or helpers) may provide more food for the young than the female, while she largely continues to brood the offspring. Keeping the body temperature of the nestlings within a safe, let alone comfortable, range is a great challenge to most avian parents, particularly those magpies living in either very hot or very cold regions of Australia.[14] The growth of downy feathers helps the young to stay warm, as does huddling together with siblings and shivering,[14] but even this is obviously not enough at first.

After two weeks of age, nestlings are often subjected to an inordinate amount of sunshine, because the nest is usually situated in a location that will get direct sunlight throughout most of the day. When the temperature is relatively high, the young begin to open their beaks and pant to keep themselves cool. In continued heat, the female parent bird will return to the nest intermittently and shade the nestlings (Fig. 8.5). She then spreads her wings and holds them slightly above the nestlings' heads and bodies. In stormy and rainy weather, and during hailstorms,

Fig. 8.5. Protection. The female responds to the panting of the nestling by spreading her wings over the offspring to afford it some shade and thus assist in cooling it.

magpie females will take the brunt of all inclement weather and sit motionless, wings spread, to protect her young. We have no reported evidence that a male parent or helper engages in this particular kind of protective behaviour.

The male will rather protect the young by keeping watch and chasing away possible intruders. If threatened by an avian predator, the female may then leave the nest and join the male in mobbing and driving the predator out of their territory. The wisdom of having helpers at the nest becomes very apparent in times of danger. The presence of helper magpies means the female can stay on the nest while the others go in pursuit of the predator or, if the helpers are young and inexperienced in raptor pursuit, it is the younger one that stays behind with the nestlings. In a pair, the female has to leave the nest and often in these situations there is a third party waiting in the wings for the absence of both parent birds so it can prey on the nestlings. In larger groups and outside the breeding season, as described earlier, the entire group will attend to any threats from predators in unison and do so in meticulously choreographed and strategically clever ways.

Physical maturation

Magpie young undergo very rapid growth after hatching. Table 8.1 shows the average ranges of weight, beak length and width, body and leg length, and claw size at five different stages of growth from a total of 36 magpie nestlings measured over

Table 8.1. Physical development of the magpie (*n* = 36).

Age	Weight (g)	Beak length (mm)	Beak width (mm)	Body length (mm)	Leg length (mm)	Claw size (mm)
1st week	50–73	22	24	51	43	21
2nd week	100–208	26	27	76	56	32
3rd week	220–250	31–38	25	89	65	35
4th week	380	52	19	104	67	35
3 months	340–400	50–55	17–24	104	71	35–37

These figures represent mean values (± 2) for beak and body measures. Range indicated when larger than ± 2. Beak and weight variations at three months may indicate sex differences (see Table 2.1, p. 22). Measurements were taken of magpies in inland New South Wales by the author.

a period of four weeks (data collected over eight years). The rate of maturity is influenced by weather conditions, food availability, genetic endowment and also the nestlings' state of health. Some of the lower weights and some low values of other developmental indices may be more atypical of healthy offspring. Beak length measured was consistently higher in wild-raised magpies as compared with hand-raised magpies (up to 5 mm difference for the same age group). Generally, at least for the New South Wales black-backed magpies, weight may increase about eightfold from hatching to fledging, which takes about a month, see 'Fledging' below. In the same period, body length doubles while legs and claws do not quite double (increase by ~75%). The ratio between beak width and length reverses. In the newly hatched, the beak is wider than longer and that ratio only changes in the third week when the beak finally lengthens and the soft edges are tightened to make the beak longer than it is wide.

Fledging

Magpie nestlings usually fledge some time towards the end of the four weeks post-hatching. Fledging involves much stress for the parents and their offspring and is a stage when many new patterns of behaviour must emerge. The young fledglings still need to be fed and remain close to the nest site, making begging vocalisations and often adopting begging postures to attract their parents.

The most important step at this stage of development is learning to fly and, just as importantly, to land. This is a vulnerable time for the young magpies because they are out of the nest but not yet able to fly off fast enough if danger is near. Coordination of muscles, assessment of flight speed, controlling flight direction, uplift and other variables are not so easy to control and only practice and experience will achieve this. Almost all young magpies do well in the actual first flying – but it is the landing (Fig. 8.6) that poses a major problem for most of them. Perfecting the art of landing requires practice and almost certainly involves learning.

Fig. 8.6. A near-disastrous landing. The point of landing ought to have been the top of the post but this inexperienced bird miscalculated. Change of posture and the action of the wings was too slow so the young fledgling slipped on the post and very nearly got itself caught in the barbed wire. With rapid wing action it managed to get airborne and then landed briefly on the ground.

The first attempts at flying often result in the young bird crashing through branches and landing in a very undignified position, caught between twigs and leaves and dangling awkwardly between branches. It shows that landing requires judgement and coordination to hold the wings in the correct position to slow flight just to the right speed in order to land on a branch. 'Overshooting' the target is a common mistake in these eager but yet awkward fledglings (as Fig. 8.6 shows).

Goshawks are ambush hunters and young magpies discovered on their own may well fall victim to the lightning speed attacks by goshawks. In one of the few studies that measured mishaps and death rates of clutches and hatched young from egg stage to six months post-fledging, there were clear attrition peaks: one peak was at egg stage for all sorts of reasons, but the major loss/attrition for their chosen species occurred within a week after fledging.[15] There are possible similarities with magpies: at egg stage, some may be infertile, or broken and the first weeks post fledging would also readily qualify as a risk period. One could add that the immediate period post dispersal may well be likely to be another peak time when young magpies may come to grief. We do not have precise data on this.

For the first two weeks (during which time they are referred to as 'branchlings') magpie fledglings practise short flights or even hop from one branch to another while gradually increasing their flight distances (see Plate 5). During that period,

Fig. 8.7. Branchling begging but not moving, being fed by the female well above the ground.

the youngsters rarely come to the ground. Instead, the adults come to them to feed them and often accompany the youngsters in flight (Fig. 8.7). Occasionally, the adults may issue short warning calls directed at the juveniles but, more often than not, youngsters seem to be rather oblivious to these.

Once the juveniles are fully flighted, they will then only be fed on the ground accompanying the parent bird, as was discussed in Chapter 5. Juveniles then stay very close to the adult and increasingly learn to discriminate food items. A set of experiments in free-ranging magpies that I conducted in the New England area (Northern Tableland, NSW) also showed that even after several months the more difficult food searches, as of scarab larvae, were largely unsuccessful.

In other words, magpie development is slow and gradual. And here is one of the seeming biological contradictions: slow development may mean several things and slowly developing organisms (called nidicolous) may ultimately have an evolutionary advantage: one of them can be that the brain not only has time to develop but be moulded by experience in such a way that it may develop further than could have been achieved in a shorter time-span. The reason for this is that the brain is nutritionally the most 'expensive' organ in the body and substantial resources need to be syphoned off into its development.

Over half a century ago, Sutter[16] noticed that during the life span, the brain of a nidicolous animal expands 8–10 times its initial size; in nidifugous animals (those that develop quickly/spend less time in their natal environment), from 1.5 to 2.5 times. So the most slowly developing vertebrates may also have the largest brains. Among vertebrates humans are among the slowest.

Reaching maturity

As magpies mature, their feather colours change slightly. Juveniles have motley, greyish feathers and even the flight feathers are not completely black. After the first moult (after one year of age), they are then equipped with the familiar colours of black and white. These changes are associated with rising levels of the sex hormones. In the subspecies *Gymnorhina tibicen tibicen* that inhabits the east coast, the iris colour changes from brown (in juveniles) to often a reddish-brown colour (compare Plate 1, adult, with Plate 4, juvenile). However, this eye colour change is not as dramatic as it is in the related currawong (brown to bright yellow).

For male magpies, physical and sexual maturity is not usually a signal for breeding but, depending on geographical location, it may be a signal to leave the parental/natal territory. The time between dispersal and breeding may involve years of fleeting and, at times, difficult social encounters, of hardships or successful bonding in relatively stable bachelor flocks. The magpie's learning experiences in the natal territory will largely determine how it will fare later on.

Survival lessons

Magpies appear to have excellent memories and we have reason to believe that they can recognise a specific magpie or human even years later. We do not know by what mechanism this occurs and how precisely tuned this is to individual features or to classes of features (such as clothes or height). Several experiments that we conducted suggest that magpies can distinguish between individual human faces and learn who is kind or hostile to them. We have also found that magpies learn very quickly how pet dogs and cats behave towards them. We found that in territories with an established cat presence, magpies tended to build nests at higher locations in the tree and also, if possible, further out on a branch than others do without the threat of agile cats.

Birds learn in many different ways. Some skills are acquired very rapidly, even after one exposure to an event or to an object or another individual. Other skills are established more slowly and only after the bird has been exposed repeatedly to the same event, object or individual. Many factors determine how long learning takes and how strong a memory will be formed.[17] Factors concerning their immediate survival are well managed and recognised in magpies. Age is important and so is the outcome of the event, such as obtaining a reward or being punished.

Learning about food

Magpies do not need to find their own food at first, since they are fed by their parents for two to three months after leaving the nest. However, the process of learning about food actually begins even during nestling stage when they may learn to recognise some aspects of food (its taste and texture in particular) and may be able to differentiate between different items. After fledging, the magpie offspring start to learn where and how to find food, how to recognise it and even how to extract it. Learning about food, as was illustrated in detail in Chapter 5, thus involves several factors.

Learning to feed on live prey generally involves several steps of trial and error. Step one is to learn to peck and use the beak effectively as a tool. Magpies use their beaks for preening as early as the first week after hatching and, while still in the nest, magpie young begin by pecking at twigs and dropping them. Step two involves pecking at a range of objects, particularly moving ones, and perfecting the art of reliably hitting the moving target. Usually this involves focusing the eyes to within the short range of binocular vision and once the bird associates eye position focus and learns to estimate distance of an object from the beak tip it becomes accurate very quickly.

Picking up food by the tip of the beak and transporting it to the back of the throat is yet another skill that is neither automatic nor all that easy to accomplish. Magpies feed their offspring by depositing the food down the throat and not at the tip of the beak. Swallowing is merely a reflex action at this point of feeding (at nestling and early post-fledging stages), triggered by the contact of beak and food in the throat. By contrast, feeding themselves involves understanding the concept of an object being edible and then learning how to manoeuvre the item from the tip of the beak to the back of the throat without losing it. Young juveniles may be presented with food scraps on the ground but then merely vocalise at it with open beaks, as if the food is meant to miraculously jump into their beaks. It takes almost three weeks before they can successfully link the sight of food on the ground with the appropriate motor activity to use the beak as a tool to pick it up and eventually swallow it.

Juvenile magpies also learn new skills through social learning from the adults. Social learning about food by observing other magpies feeding is important throughout life. Very complex manoeuvres can be learnt by one bird watching another bird performing them. For example, magpie fledglings walk close to an adult most of the time and learn by watching and listening. As is explained in Chapter 5, the biggest challenge for magpie juveniles is to understand how a sudden dive by a parent's beak into the ground produces edible food. Extractive foraging is considered to be one of the most complex feeding tasks, as mentioned before, and this is largely limited to some avian, canid and primate species. The juvenile has to learn to hear the minute sounds of larvae moving under the surface first and then learn to associate these sounds with the action of the adult magpie and the procurement of food. This is a long and complex learning process. Survival depends not only on pecking food but also on avoiding inedible objects and poisonous insects. The learning has to be fast and accurate but there have been as yet no experiments to show how magpies perform such important discrimination tasks.

In the final stages of learning about food, the juveniles are slowly 'weaned'. That is to say the adults will increasingly withhold food and feed themselves in front of the 'outraged' juvenile. It is then that we hear the particularly loud and persistent begging calls of the juveniles who do not seem to tire in their requests for food but, increasingly, find such requests ignored or even penalised.

Learning about predators and risks

Adult magpies will mob eagles, owls and other birds of prey, such as hawks and falcons of any kind, and this involves flying around the potential predator while making loud vocalisations. It is reasonable to assume that, at some stage during their development, young magpies may have the opportunity to see how adult birds deal with birds of prey and thus learn that these are species to be feared and mobbed. However, it is possible that, like other bird species, they have innate fear responses to specific physical features of birds of prey, such as splayed primary feathers or the size of birds of prey. Studies of the responses of juvenile magpies to predators have shown conflicting behaviours. One study suggests that mobbing of other predators has to be learnt by observing other birds;[18] on the other hand, there is at least one reported case of a New Zealand hand-raised magpie that suggests that a response to a snake was spontaneous and not learnt.[19] In a study we conducted in the field, we found, however, that there was a distinct behavioural difference between naive juvenile magpies and adults. We could assume that the new crop of juveniles was likely to be 'naive', i.e. had never seen a snake before, because the weather was still consistently cold (down to 5°C at night and barely above 12°C during the day) and that this was too early in the season for snakes to appear above ground. We attached a rubber snake of indistinct dark colour, mimicking young tiger and copper snakes, to a long nylon thread, placed it in an area to which magpies preferentially came mid-morning, rolled it out to 5 m and then hid behind a natural visual barrier. When the birds arrived and were in visual distance to the snake, we very slowly wound the nylon thread in, mimicking snake movement. Adults stopped foraging as soon as they saw the snake, while juveniles tended to approach the moving fake snake and showed signs of pecking at the stimulus rather than vocalising or showing any fear responses. Even when the snake had been entirely removed, the adults did not resume foraging but remained in a vigilant position, eyes scanning the ground for another three minutes at least, and then chose to fly a short distance away before foraging again. The results were significant.

On each trial the juveniles reacted by approaching the snake, showing curiosity without the slightest body posture change. One juvenile actually pecked the snake. By contrast, the adults were at once alert showing this by body posture while in some trials a short alarm call was issued.[20] It seemed obvious that the juveniles had no idea that snakes posed a real risk and the juveniles showed no innate neophobia.

The adults' sudden change from foraging to vigilance behaviour and, in some cases, the brief alarm calls directed at the juveniles could well have been enough for the juveniles to have learned to avoid snakes in future. Indeed, magpie adults (parents and helpers alike) appear to spend a good deal of time in communicating danger or risk to juveniles. The level of danger or risk is communicated by different alarm calls that differ in decibels and sound structure (see Chapter 11 for more detail) and by appropriate action, such as alarm calling first and then flying off while calling the juveniles to follow. In many instances, the adults will interrupt foraging, actually turn their heads in the direction of the juveniles, suggesting intentionality, and emit a series of calls until the juvenile has responded.[21]

There may also be occasions when teaching via vocalisations may be achieved by a direct method associated with feeding. For instance, magpie parents may vocalise just before feeding their nestlings or fledglings, i.e. while the food was held in their beaks. Some such vocalisations were mimicked sounds of other animals in their territory. These typically did not belong to predators but to species, such as horses, that pose no direct threat to magpies, as if to distinguish between risks and non-risks.

Vocal learning

As the magpie is a songbird and very impressive in its song, range and vocabulary, a much broader discussion of its remarkable abilities is certainly needed. It is no exaggeration to claim that the magpie is one of the foremost songbirds in the world. Vocal learning is thus a key topic for this extraordinary bird that deserves far more elaboration (see Chapter 10). A very intriguing question is how magpie vocalisations are learnt and how much they are innate. It used to be thought that species-specific song is innate, but there is now overwhelming evidence that many species of songbirds learn their songs by copying model songs heard early in their lives.[22,23] In many oscine species studied so far, learning of song is usually confined to a limited period of life, i.e. a sensitive period,[24,25] and in most species that period is in the first months of life.[26] Thus, many birds learn their songs as juveniles long before they ever sing them themselves. They are silent when learning the songs by listening to other birds singing and only later do they produce the vocal repertoire that they have learnt. This is not the case in magpies.

Indeed, the models of songbird learning that have been elaborated over decades and are based on very careful and detailed studies do not seem to fit all Australian native species. The preferred models, songbirds of the Northern Hemisphere, as well as the zebra finch that has similar song learning habits as songbirds at high latitudes, simply do not match the theoretical models. In fact, they sometimes make relatively little sense in the Australian context.

We now know that all modern songbirds (and many other bird families) actually evolved in Australia before they radiated out to other continents and we know that the species we have on the Australian continent are the founders of all modern lineages worldwide. This means that we no longer have to try to somehow fit our species into established models, but instead can ask important questions about the evolution of song and of bird behaviour generally and then proceed to ask what might have changed and why since songbirds left Australia.[27]

A particularly intriguing finding of my own research was that juvenile magpies practise song while still in the nest but they do so exclusively in the *absence* of adults. In over four years of observing nesting magpies in local territories, magpie nestlings persistently stopped vocalising when the adults came within 20 m of the nest.[28] While this does not diminish the possibility that adult birds act as tutors, their social organisation suggests that adults are not appointed tutors, as in zebra finches for instance,[29] and that vocal learning at least is not an activity between adult and offspring. Indeed, vocalisation by nestling and juvenile magpies is suppressed in the *presence* of adults. The only calls of nestlings and juveniles up to six months of age ever recorded in the presence of adults are begging, distress and brief alarm calls.[30]

There is yet another level of learning in which magpies have to excel, and a highly important one, and that concerns navigating the social environment – the subject of the next chapter. Indeed, learning about social 'etiquette' in magpies, about group hierarchy, neighbours and strangers can be a matter of life and death. Social learning also influences vocal learning that, in turn, is needed for effective communication.[31] This suggests that magpie society is a complex social society, a topic to be discussed fully in the next chapter.

Dispersal

It has been customary, based largely on banding records, to suggest that magpies disperse in their first year but they do not disperse far from their natal territory. Such banding records have revealed dispersal distances usually under 20 km[32,33] and more often than not no more than a few kilometres. While this general information might hold overall, there are many exceptions to this rule. One exception is that magpies may, in fact, disperse over long distances and another, at the opposite end of the scale, that they may not leave their natal territory at all. In fact, not all magpies disperse in a given season.

A study by Veltman and Carrick[34] examined the dispersal patterns based on re-sightings of banded birds over a 12-year period (1953–1965). It showed that

12% of juveniles did not disperse at all while others left the natal territory only after two, three or four years. There appear to be advantages in not dispersing. Their records show that non-dispersing birds suffer a substantially lower death rate than dispersing juveniles. In a 12-year period, only two birds were killed and both cases were deaths by human hand or human technology (one was shot and the other electrocuted).

Dispersal age may thus vary as well. Dispersal age is usually between 8–10 months of age. In some extreme cases parent birds start harassing their offspring to leave when they are just six months old. In these extreme cases, the juveniles are usually driven out within a month but spend the last month in their natal territory under difficult conditions, being barely tolerated by the parent birds. At the other end of the scale, there are no departures at all and the young are allowed to stay and may do so for years or even for the rest of their lives.

Sex differences also have an effect on the dispersal age and dispersal patterns. In the Canberra area, Veltman and Carrick found that, in a given cohort of offspring, males dispersed at an earlier age than females, regardless of whether they dispersed in the first, second, third or fourth year and the cumulative percentage dispersals were higher for males than for females in any given departure year.[34] The dispersed males were found not to settle with kin. Some of these patterns are unusual when compared to other passerine birds, particularly group-living birds. As a rule of thumb, males are philopatric (i.e. they tend to remain near or return to the same location) while females disperse further and settle in different localities.[35] In magpies, however, it appears that females are more philopatric while males disperse, although some males may stay in their natal environment.[36]

Long-distance dispersal

Magpies are often perceived as having considerable numerical and locational stability. However, BirdLife Australia, the Australian ornithological umbrella organisation, has successfully run Australia-wide counts of birds and has accumulated vast records of numbers of birds per season in given geographic locations. Interestingly, distribution of magpies across Australia between seasons of the year (Fig. 8.8) shows substantial fluctuations and, contrary to general belief, high density of magpies is found in areas of the lowest density of human populations. In general, the data analysed reveal that there is considerable movement of birds, classified as migratory or not, in all directions, most notably are those from coast to inland during the winter months (Fig. 8.8).

It appears that magpies are less numerous along the entire stretch of the east coast of Australia than in the inland. We are unfortunately not in the position to compare congregation densities of magpies of the 18th, 19th and 20th centuries because the data are simply not available, but it is conceivable that European settlement has effectively changed dispersal patterns. It may be that more magpies

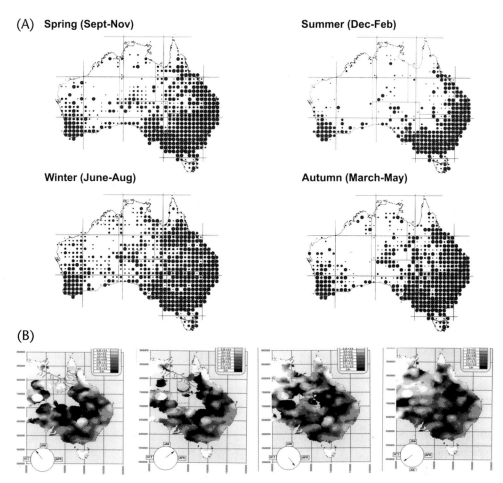

Fig. 8.8. Seasonal variations and concentrations of magpies. A, Method 1: Recordings of magpie concentrations across all staffed/volunteer transect recordings (reproduced from Barrett et al. 2003[37] with permission from BirdLife Australia. B, Method 2: Computer-generated series of maps showing just some of the enormous fluctuations of magpie movements over the seasons. The darkest marks show the highest concentrations; blank areas represent no magpie presence. The driest and hottest parts of Australia have no magpie presence between October and January (panel 1) and January to April (panel 2). Between April and July (panel 3) and July and October (panel 4) (i.e. cooler and winter months), magpies have been recorded everywhere, even in dry weather. (Reproduced from Griffioen and Clarke[38] with permission.)

die in coastal areas rather than moving inland, but it may also be that more magpies actively move inland and north.

The next unexpected finding concerns the dispersal patterns. Substantial shifts in density of Australian magpie populations make no sense unless one considers the possibility that part of the new crop of dispersing magpies is actually engaged in much longer-distance dispersal flights than has been thought or that there is an exceptionally high death rate of young birds in some areas. Alternatively, magpies in inland Australia and at the Top End may be seasonally nomadic, driven away in summer by the heat in the arid inland and by the wet in the Top End.

It has long been observed that, in Australia, there are no clear distinctions between non-migratory and migratory species.[39] There is a gradation from those in which the entire population commutes seasonally between breeding and non-breeding locations and those of entirely sedentary habit – intermediate species show every degree of partial migration, dispersal and nomadism. Even regional populations of the same species may have different seasonal movement patterns and, to an extent, this may be true of magpies.

Short-distance dispersal

Dispersal patterns over short distances are quite complex. While distances covered at specific dispersals are usually between 3 km and 20 km, the question of who disperses, how often short dispersals take place (of one individual over a lifetime) and whether individual magpies eventually settle in a place where they originally dispersed to is not so easy to answer. Generally, we speak of dispersing offspring before the new breeding season begins and of adults as sedentary. However, in the south-western region, exchange of females per generation is the lowest of all measured across Australia[40] showing that not all juveniles disperse immediately (delayed dispersal especially by female offspring) and some may never leave while others, in other regions, may take up a semi-nomadic lifestyle for many years of their adult life.

The picture that has begun to emerge for magpies across Australia is that many adult magpies also remain mobile and continue to disperse to other areas and may do so several times during their lifetime. Such adult mobility may have a variety of reasons. For instance, as already described, various groups (marginal, mobile, open groups and flocks) share instability as their chief characteristic; many of them, as Carrick noted, lasting for less than a year.[38,39] There are constant changes of boundaries and affiliations and thus also of location. Flock birds are locally nomadic and so are all other birds that live in any of these less stable groups. While the word 'sedentary' best describes the ideal life-condition for adult magpies – those with permanent residency status – there is considerable mobility for the rest of the adult population and thus magpies have a dynamic system rather than a static one, reflected both in their social complexity and in shifts of territory.

As will be described in more detail in the coming chapters, such variability in rights of occupancy and resulting mobility also deeply affects bonds and group structures. Suffice it to say here that the magpie's patterns of group life and territorial shifts have few equals among avian species. The only semblance to their complicated group structure and variability that I have been able to find is that found in the carrion crows of Spain, *Corvus corone corone*.[41] Their social groups show a similar mix of monogamous and unassisted pairs, some polygamy, some groups with helpers and some communally breeding groups including offspring that had not left the natal territory. As do magpies, these crows also constitute groups, but not always, and they do so by either retaining females or by accepting male immigrants into existing groups and by long and short-distance dispersals. Unlike white-winged choughs,[42] superb fairy-wrens[43] and many other Australian cooperative breeders, magpies are not always cooperative (as described in Chapter 7) but their social system is flexible enough for re-arrangements and some of this flexibility is evident in variability of group sizes and in dispersal patterns.

Reintroduction of nestlings/juveniles

A very special case of tolerance of a 'stranger' occurs when a nestling that has been taken into human care because it was injured or had fallen from its nest is reintroduced to its natal territory. Within the relatively small magpie groups of G. t. tibicen that I have observed on the New England Tableland, there seem to be definitive time limits for such reintroductions. Over a period of eight years, I re-introduced 22 fledgling magpies to their natal territory and noted the responses of the parent birds to the returning youngsters by period of time of absence (a few days, a week, two weeks and even three weeks).

Of the 22 fledgling magpies, half were returned to the parents within a week. In these reunions the adult response was uniformly positive. The released bird flew straight towards the female parent, accompanied by loud vocalisations. Once the young bird had reached the parent bird it immediately assumed a submissive posture, used a begging call and sometimes also fluttered with the outside primary feathers, wings held apart. It is not clear whether this behaviour is some kind of greeting display or an added measure of appeasement. To date, no such display has been described for magpies. Greeting displays usually have either one of two functions: to re-establish close bonds (as in albatross bonded pairs meeting after some time apart) or, as a more casual and distant encounter, merely to establish the individual's identity within a colony or larger group as that of belonging rather than intruding.[44] It is more likely to be the former, or it could be a signal to allow the adult female to identify the chick as hers. This is not known. It may be that all these reasons apply. Of interest is that such reunions are almost played out identically in different groups. All other magpies in the group quickly assemble in the same spot and usually inspect the returning youngster. The others then fly off and the parent bird resumes foraging with the reunited offspring following.

The other half of the young birds (often received as nestlings and returned as fledglings) could only be returned two or even three weeks after their rescue because of the nature of the injuries or problems. Those that were returned two weeks after separation were still welcomed by the parent birds in a similar manner but the remainder of the magpie group tended to show some initial agonistic behaviour, carolling and adopting threatening postures. It depended very much on the behaviour of the young bird whether such agonistic postures extended to the adult group members administering some painful jabs with their beaks. If the fledgling showed submissive behaviour and continued with begging sounds, the antagonism soon waned and the juvenile was fully accepted back into the group.

When the birds were reintroduced at a more advanced age, after three weeks of separation (as was unavoidable if they had suffered a fractured bone), clear signs of friction were evident by the low level of tolerance that the rest of the group was prepared to give. My observations show that, in these cases, the young bird may be lucky to get the support of a parent bird. It will move to the bottom of the hierarchy and will face a life of severe surveillance until ready to disperse from its natal territory.

We do not know whether individuals recognise each other after long periods of separation or, for that matter, whether 'absence' in dependent offspring, especially if for several weeks, signals dispersal and a 'never to return' departure.

Whatever the explanation, a two-to-three-week window of opportunity for a fledgling or juvenile to return safely to the parent birds and be accepted back is still remarkable behaviour. In kookaburras, returning offspring tend to be rejected after a week to 10 days (if other offspring are present); in wattlebirds, the maximum period of absence is three days – any later, and the very belligerent wattlebird parents will launch a full attack (with intent to kill) against their offspring, treating it as an intruder.[45]

Endnotes

1. https://www.theguardian.com/environment/2017/dec/11/magpie-edges-out-white-ibis-and-kookaburra-as-australian-bird-of-the-year
2. O'Connor 1984
3. Potti 1999
4. Ricklefs and Starck 1998
5. Williams 1994
6. Potti and Merino 1996
7. Rogers 1995
8. Vince 1973
9. Teather 1992
10. Kaplan 2018a
11. Larkins 1980
12. Kaplan 2017a
13. pers. observation
14. Visser 1998
15. Tarwater *et al.* 2011
16. Sutter 1951
17. Clayton and Soha 1999
18. Curio 1988
19. Brockie and Sorensen 1998
20. Koboroff and Kaplan 2006
21. pers. observation
22. Kroodsma *et al.* 1982
23. Bell *et al.* 2014
24. Schneider and Mooney 2015
25. Vallentin *et al.* 2016
26. Marler 1991
27. Cracraft *et al.* 2004; see also Kaplan 2015
28. Kaplan 2000
29. Slater 1989
30. Kaplan 2018c
31. Hausberger 1997
32. Baker *et al.* 2001
33. Jones 2002

34 Veltman and Carrick 1990
35 Brown 1987
36 Hughes *et al.* 2003
37 Barrett *et al.* 2003
38 Griffioen and Clarke 2002
39 Carrick 1972
40 Baker *et al.* 2000
41 Baglione *et al.* 2002
42 Boland *et al.* 1997
43 Double and Cockburn 2000
44 Kaplan and Rogers 2001
45 pers. observation

9
Social rules and daily life

What does daily life look like for a magpie? When thinking about the social life of magpies, one might think of John Donne's famous line 'No man is an island': there are no magpie loners at all. No matter how imperfect the situation may be, any social contact is better than none. Magpies come in friendship bundles, as temporary travel mates, as travelling groups, couples and family birds, be this nuclear or extended family. If there is no magpie to be had as companion, they will form unusual friendships, the best-known are with people's pet dogs and sometimes even with people. And some of the associations that magpies form are not just superficial, pragmatic or 'marriages of convenience'. The unusual quality about magpies is their voluntary association with others and the dedication with which they adhere to such associations even when these are not the norm. Of course, many people have pet birds and often have close and even affectionate relationships with them, largely with parrots, but this is qualitatively not the same, as I hope this chapter, among others, will show.

A life in the constant company of others has its advantages and disadvantages and it can even depend on the time of year where the pendulum might swing. One of the many misconceptions we have, embedded also firmly in verbal expressions (such as 'behaving like an animal'), is that 'wild' means rule-free. Nothing could be further from the truth. Behaviour patterns that are part of group life and not related to territorial defence show a great deal of variation and complexity but they are governed by a set of strict social rules. Any defiance of these rules (and this is the only way we can observe that the rules exist in the first place) is usually followed by swift punitive actions or appeasement gestures.

Daily life and activities

The most leisurely months in the year for magpies are probably around December to April, when the offspring have been raised and may now find their own food and food is usually plentiful (depending on local conditions and on the absence or presence of dramatic climatic events such as bushfires and hurricanes). At this time of year, magpies commonly finish their first feeding rounds by mid-morning and usually do not need to resume their walk-foraging routines for several hours. There is no other way to describe it – if there are no urgent matters to attend to, such as trouble at the border, a bird of prey or a cat to worry about, these magpies have actual leisure time! And how do they fill it?

Health care

One way is to indulge in personal health care and magpies can spend an inordinate amount of time pampering themselves. Sunbathing is equally a popular activity but this is believed to have direct health benefits and is less likely to figure as a leisure activity. Adult magpies sometimes lie prostrate on the ground and do not respond to anyone approaching – as if in a trance. Concerned individuals sometimes approach, wondering whether the bird has been injured and on reaching the magpie, it suddenly springs back to life. There is little doubt that this is an enjoyable experience for the birds when the sun can penetrate to their skin, possibly dislodging any ectoparasites (Fig. 9.1).

Fig. 9.1. Sunbathing. A, The adult is still standing but all feathers are splayed and the head turned to maximise exposure to the sun. B, The nestling is lying on its back and also tilting its head to get sun exposure to the neck – note that the nestling has not closed its eye.

Engagement with water and playing with water is enjoyed by young and old. Danielle, 'the magpie whisperer' (www.magpieaholic.com), has set up her entire backyard as a spa bath and playground for magpies. Magpies are extremely fond of water, take a bath every day if the opportunity is there (or even two dips per day) and then spend a good deal of time preening post-bathing. The birds have extremely clean plumage. When a gentle ground sprinkler is set up, the games can begin – running into the spray, running through it, flying just above it and then dropping down to get their bellies wet and sometimes snapping at water and making little sounds (seemingly in pleasure) to accompany the proceedings (Plate 7).

Singing

Apart from extended attention to sun and water, as a rule of thumb, one of the most remarkable and sustained activities at times of leisure is singing. For anyone with some exposure to magpie song this would be a comment perhaps not even worth making. However, in the literature of birdsong it certainly is. Birdsong is meant to have a function. Song ought to be either for reproductive purposes (males singing to attract a female) or for territorial defence. Well, the magpie breaks all rules! Magpie songs not only have no obvious function, at least none has been found to date, but song is not even sex specific, i.e. not a prerogative of the male.

Males and females sing alike and they do so to an extraordinary degree, especially outside the breeding seasons. I took the trouble to measure all warbling/song activities through the year and over a three-year period in different territories. I also recorded all other vocalisations (for a complete description of the annual cycle see Kaplan).[1] Song in magpies is particularly interesting because it sharply drops off in the pre-breeding season, around June (Fig. 9.2).

In songbird species of the Northern Hemisphere most song is produced in the pre-breeding season. In the magpie, song almost disappears pre- and during the breeding season.

Even more interesting is the length of time of such vocalisations. Fig. 9.2 counts only the onset of a bout of warbling. It does not tell us anything about the length. Hence, a different measure was taken (over one season only) and, when read together with the above figure it shows that the air is filled with the tuneful songs of magpies blanketing the soundscape of their territories for most of the day (Fig. 9.3). Many of these songs are not just extremely long but continuous, like a perpetuum mobile. The longest continuous song I have ever recorded was 6 hours long, without interruption or break.

This soft song at low energy and decibels (46–65 dB) is characterised by three distinct elements: (1) it requires no particular body posture: it can be performed sitting or standing in a relaxed posture, therefore there are no postural demands as there are in carolling (of that later); (2) it can be performed as a continuous sound;

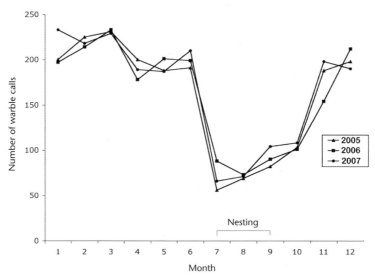

Fig. 9.2. Annual variation in quiet song output, averaged for three territories. Note the rapid and substantial drop in singing before the breeding season, the very opposite of song activities around breeding time in high latitude bird environments (see also Plate 6 for another example of quiet song performance, with beak half open).

and (3) usually song is produced in the syrinx but in this type of song the larynx and the beak in particular are likely to play an important role in modulating the sound. To produce the song, the beak was either closed completely during vocal production or opened just a fraction and for very brief periods only (Fig. 9.4).

Maintaining a closed beak minimises air loss on expiration. The singer, by taking mini-pauses, can continue to produce ongoing vocalisations like a

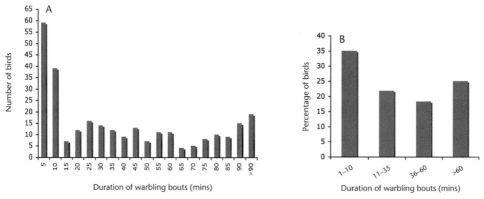

Fig. 9.3. Song duration in magpies ($n = 146$). A, The number of birds recorded warbling for various lengths of time. Note that song duration varies considerably, from 5 minutes through to over an hour and a half. B, It was most common for birds to warble for 1–10 minutes in one bout of song; however, there was a fairly even spread across four time periods. The variations of length of warbles between and within individual birds were substantial. Some individuals warbled regularly for no more than 5–10 minutes at a time while others engaged in song bouts of more than 1 hour or alternated between short and long song sessions.

Fig. 9.4. Quiet song with closed beak. The arrow points at the larynx, indicating that the feathers near the larynx are moving while song is performed, an energy saving practice that allows the magpie to sing continuously for hours (see also Plate 6 for another example of quiet song performance, with beak half open).

perpetuum mobile. Warbling is a low energy but complex form of song and suggests a similarity with 'humming' in a human voice may not be far-fetched. Very little energy is needed for humming and warbling may be similar in that very little air has to be expelled to produce sound and no other musculature is needed as in high amplitude vocalisation. Low intensity should require less energy and rather low metabolic cost, as was found by Jurisevic[2] and hence the description of a leisure activity.

I also collected data of the time of day for such singing activity because it is tempting to assume that such singing might take place early in the morning, as is common practice in several songbirds (dawn chorus). Magpies do not do this. The data were taken from three different territories in New South Wales in the months of January and February. As Fig. 9.5 shows, singing takes place during the day and peaks around midday and then slowly declines in number and amount. This means that at those times when most other birds are silent magpies have the stage more or less to themselves.

When magpies sing their quiet song they tend to keep by themselves as if to allow each other space for the performance. They also tend to have a nap after such a solo bout and then may complete their lunchtime program with some preening activity – all the while not moving from their chosen spot. More will be said about song and communication in Chapters 10 and 11. Suffice it to say here that the melodious song is for the singer, not for an audience or any obvious purpose. Of course, for any human eavesdroppers it is a free concert of the nicest kind.

Before dismissing the possibility that the magpie's song may be regarded as a leisure activity, in species so far studied, it is interesting to note that there is

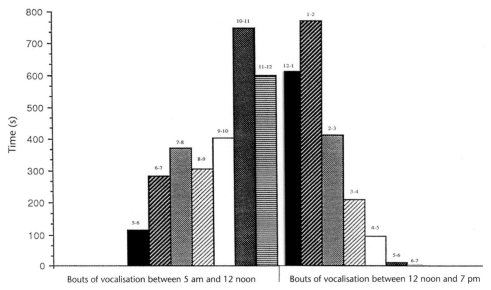

Fig. 9.5. Vocalisations across the day (Jan–Feb averages). The actual bouts per individual could be much longer on a given day but as a proportion of singing activity it seems clear that most singing takes place between 10 am and 2 pm (25 min (30%) morning, 45 min (55%) between 10 am and 2 pm, and ~12 mins (14%) all afternoon and evening).

evidence of singing-regulated dopamine[3] and auditorily stimulated release of dopamine and serotonin in the bird's brain.[4] Dopamine is the 'reward' neurotransmitter and serotonin alleviates stress and promotes relaxation and sleep. In other words, singing may well be a pleasurable experience, even induce euphoria in the singing magpie. This could certainly be a motivator and the function of song in magpies might even be described as inducing a sense of wellbeing and low stress levels, factors that certainly contribute to survival and long life.

Group activities apart from feeding and defence

In some magpie groups, subgroups and friendships form and such groups sometimes go walkabout. They may fly a little or actually proceed on foot without any evidence of food searches. In some of the meetings, one almost has the impression that the magpies are talking to each other (Fig. 9.6) and, of course, there usually is some close-range vocalisation, something we have not yet deciphered.

Magpies, unlike parrots and tawny frogmouths, are not a cuddling species and don't roost by touching each other. However, alliances as the one shown in Fig. 9.6 indicate that these associations are held together by a strong sense of belonging and sharing the duties and pleasures of its territory.

Sometimes, there are mother and daughter alliances that are measurable in the sense that one can count the times they spend together on a daily basis, assess the evidence of approaching each other and do things jointly. For instance, a pair had

9 – Social rules and daily life | 145

Fig. 9.6. Time off for residents. Close allies spend some time with each other. (Photo: Bronwyn Gould)

already been in the same territory for seven years and had always ensured that their offspring left before the next breeding season. In the eighth year, the female's relationship to her daughter (the sex of the offspring could only be determined after the first moult when the bird had its first adult plumage) was very obviously different. The two were inseparable and could be seen spending most days together, well beyond the time when this offspring needed parental assistance (Fig. 9.7). At the time this image was taken, the other two offspring had long since left the natal territory. Leaves had dropped and the first signs of winter were in the air when it became clear that, unlike previous practice, this daughter could stay on and, even more surprisingly, she was allowed to stay for yet another year thereafter, giving this particular offspring a significant head start in life.

Life is lived in the fast lane when territories begin to close up in preparation for the breeding season, the focus for magpie life shifts. However, during the good times with low levels of responsibility it is time for fun – and even for play.

Play behaviour

One of the 'leisure' time activities that are most pronounced in the months from October to June is play behaviour. Juveniles may play extensively, especially if they

Fig. 9.7. Mother-daughter team, adult female on the left and the juvenile on the right (note the evenly scalloped feathers on chest and abdomen typical of juvenile plumage). Watching them, one got the impression that the daughter mimicked her mother's behaviour. She was looking where her mother was looking, flying to where the mother had landed and even some micro-behaviours such as preening seemed mimicked because it typically followed on directly from the mother's preening.

no longer need to be fed exclusively by an adult. Adults may even engage in some limited play behaviour themselves but the repertoire of juveniles seems endless and most delightful. Adults tend not to partake in social games but may engage in some acrobatics, such as hanging upside down and swinging (Fig. 9.8). One adult discovered our Hills Hoist washing line and was hanging from one of the clotheslines and I discovered that I could push the hoist around while the bird was hanging on. It became a game we played and, evidently, the bird enjoyed the free rides and regularly came back for more.

Many mammals play. Usually, in cognitive research, three forms of play are identified, ranging from simple forms to very complex forms of play that require some thought, insight or understanding of concepts or principles.[5]

The simplest form of play consists of dangling, running, i.e. self-involved play that needs no other parties or objects. The second category involves objects that

Fig. 9.8. Hanging and swinging, not very good when the anchoring point is a piece of clothing or a towel but great fun for the magpie youngsters when they can swing on the Hill's Hoist. (Photo: The Magpie Whisperer, www.magpieaholic.com)

can be carried in the beak, such as leaves, twigs, stones, and occasionally even man-made objects such as pegs, especially colourful ones, or as Fig. 5.2 shows, even discarded bits of children's toys (see also Plate 9). They may be carried, dropped from flight, picked up again and again. Elsewhere, I have described such behaviour in kookaburras and black kites and, of course, bowerbirds have a particularly strong penchant for colourful objects.[6] The third category is called social play, needing at least two players but will often have more and may involve several games that human children will also play spontaneously.

The literature has tried to categorise these three forms of play (solo, object, social) to ascertain how many species worldwide are engaged in each form and whether there are different cognitive requirements. Indeed, solo play is shared by many species worldwide but the number of species involved in play behaviour drops off progressively from one category of play to the next. In this last category of social play so far only relatively few avian species have been found to clearly display social play behaviour. Overall, among vertebrates generally, social player numbers dwindle and may probably account for no more than 1% of all vertebrates.[5]

Birds in the category of this most complex form of play typically engage also in the two other, simpler, forms of play but, conversely, there is no guarantee that birds engaged in solo and object play ever play socially. Magpies belong to only a very select number of avian species so far known that engage in extensive play behaviour in all three categories as juveniles and even into adulthood. Parrots and cockatoos have an extensive social and individual play repertoire but there are few other avian species, apart from some corvids, with these traits.[7]

In other words, for a songbird, magpie play behaviour[8] is highly unusual and has only been matched and confirmed in common ravens, a species that is considered cognitively highly advanced. It is akin to parrot behaviour and, indeed, even to that of dog pups or young primates, involving social play with another juvenile or an adult and individual play. They roll around, breast bump each other, pull on wings, run after each other, hide, crouch, engage each other with legs, play fight, peck, grasp, jump and even play hide-and-seek.[6] Magpies are also insatiably curious (see Plate 12).

Object and social manipulation in magpies begins at about four weeks after fledging.[9] Juveniles as well as adults may hang upside down, dangle and swing, even just by one foot (Fig. 9.8). They also manipulate objects with their feet, as we generally see only in primates and some parrots, and will even play by themselves with objects, such as sticks. If an object is obtained, the juvenile may run away with it as if it is a trophy (as shown in Fig. 5.2). These play bouts have been described as lasting for several minutes[10] but they may, in fact, be much longer – I have recorded play sessions that have lasted up to 10 minutes.

In magpies, social play is the dominant form of play (see Plate 8). Juveniles will even invite each other to play. And within this category, hide-and-seek and play fighting are the most common games.

More unusually, magpies may engage in cross-species play. Hand-raised magpies may playfully engage with human adults. Playing games with shoelaces is a favourite sport and these get 'attacked' and pulled. Probably the most famous case is a YouTube video of a magpie playing with a Jack Russell and apparently doing so on a regular basis.[48]

Play fighting

When the juveniles get a little older and are nearly ready to leave the natal territory, they start engaging in rough and tumble play-fights that can take many hours. I once watched a group playing all afternoon. The reason for the sustained interest was probably the fact that this was a meeting point of three different territories (Fig. 9.9). Within one territory and only a limited number of potential players the games probably would not last quite as long. The sparring with playmates from other groups may have held special interest and kept the birds motivated.

Fig. 9.9. Play-fighting of juvenile neighbours. A, juveniles fly in from various places and the game starts. B, between two and five participants join the game, pulling tail feathers (or trying to) and making frontal attacks. C, the intimidation game: wings spread out wide. D, declared victory of magpie with outstretched wings looking at the vanquished opponent who has already admitted defeat by being on its back. The commotion is enormous and the defeated magpie on its back screeches. Then the game starts all over again.

Like teenage kids hanging out at a street corner, the novelty of interacting with others not from their own family and the sparring of strength and skill kept the magpies totally engrossed. Mostly, individuals would not jump more than a metre high so almost all play-fighting took place on or near the ground. It did involve speed and quick wing manoeuvring, however. The point of the play, as far as one could tell, was to subdue one another, but stop at the moment when the fight could become physical and inflict pain.

Friction between breeding females

Aside from play, there can be substantial internal friction within a group that seems to be more pronounced during the breeding season. If the group consists of more than one breeding female, dominance patterns may emerge and may lead to intense harassment of the subordinate breeding female by the dominant one. Even though the subordinate female may be a daughter of the dominant one,[11] the dominant female may not rest until the eggs of the subordinate are destroyed. She may even destroy the nest while it is being built, as mentioned in Chapter 5. Alternatively, the dominant female may employ the subordinate female to brood her own eggs. In resource-rich years, however, subordinate females may be allowed to raise a brood, leading to an increase in the total number of viable offspring within the group.[11,12] In some regions, plural breeding (more than one female nesting) may be relatively common without causing any friction.[13] The latest findings even suggest that extra-pair matings in magpies are extremely high[14] and this may involve a good deal of rule breaking but not necessarily lead to detection or friction.

In-group friction

Frictions and stresses may not always be avoidable but the question is whether and how they are resolved. Over the years, we have uncovered a complex social system and rules to maintain peace and harmony and keep the group safe. Similarly, as happens in dog packs, it surprised me most that punishment was very swift, apologies were quickly received and the recipients would, at the end of the conflict, accept the individual back on the same terms as before the conflict and calm would return in an instant. Interactions of this kind tended to take less than a minute yet they held the social fabric together.

It seems, from long-term observation, that apart from very small skirmishes, occasional flare-ups of friction within the group have not so much to do with territory as with social status, hierarchy and even 'trust'. It is possible to find magpies within a group expressing agonistic responses to a long-standing member of a group or towards an incoming unknown magpie. Such a magpie may not

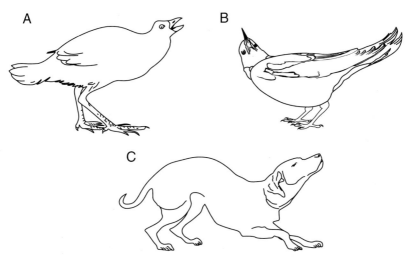

Fig. 9.10. Crouching, a well-known posture of submissiveness found in many animals: A, Tasmanian native-hen; B, magpie; and C, dogs.

necessarily be permitted to feed together with the group or roost with them but may be allowed to stay and participate in the group's defence of the territory.

Conflicts can range from conflicts over short-term issues to general misbehaviour that could harm the group as a whole. A short-term personal conflict between two magpies could be that both discovered a food source and the one further away flies in and snaps it away in front of the other.

If thievery is involved, the result is an immediate attack on the thief and, just as immediate, an apology by the thief. Such an apology is expressed sometimes in the form of accompanying vocalisations but always by adopting a submission posture.

There are at least three types of submissive postures. These are shown in Figs 9.10–9.11. The first figure refers to a well-known crouching position, first reported and made famous by Charles Darwin who noticed such a posture in dogs. The other two submissive postures are in much more benign contexts, namely chiding offspring.

What are misdemeanours in magpie society? For juveniles, the rules are quite Victorian: offspring must be seen, not heard. Youngsters get 'disciplined' by pecks if they break a social rule,[15] as do adults.

To fly away and not come back when called is a misdemeanour. To vocalise the wrong type of call may be a misdemeanour. Juveniles are allowed to produce begging calls and distress calls – both will receive support from adult birds. However, alarm calls, unless brief involuntary expressions of fear or through being startled, are also not permitted. A severe misdemeanour is to attempt to carol, i.e. using the territorial call.[49] That is solely the province of the adult resident magpies. Any attempt to do so will be severely punished. In those cases, the offending

Fig. 9.11. Representation of the various forms of saying sorry. Both A and B are variations of the submissive posture. In A, the juvenile crouches in front of the adult male (likely to be its father) but chances a glance at him. Whatever this means as a gesture, it works. The male stood there for a while but did not punish and eventually left. The juvenile was allowed to get up and resume activities. In B, two females vocally 'comment' in detail on the behaviour of the juvenile and the juvenile stays put motionlessly, eyes averted. Females rarely physically punish their offspring and the kind of submissiveness appeased the females.

magpie will roll onto or throw itself on its back, legs moving in the air as a limp attempt to protect its exposed belly, a posture also used in play behaviour (see Fig. 9.9D). Another set of rules involves food and who can eat what and when. Juveniles are not allowed to help themselves unless they have asked and are certainly not permitted to take food that is in front of an adult unless specific permission is given and expressed, for instance, by the adult turning away. The latter is part of the advanced training of the young to feed themselves. Excessive begging and blocking the way for the feeding adult, mostly the female at the latter stage, is tolerated in two-month old fledglings but not in those three-months post fledging.

The severest cases of perceived wrongdoings, usually not found in juveniles, apparently do not involve submissive postures as described above but 'hanging heads'. A bird stands upright but hangs its head forward offering the top of the head to be pecked with body visibly shaking. One of my hand-raised magpies that I released together with other hand-raised ones continued to stay with his original group but, after a few weeks, I found him severely pecked on the head and bleeding from other parts of his body. I took him back into care and later re-released him. He promptly flew back to the very group that had so severely punished him. Several months later, after the feathers on his head had finally grown back, I once again found him, beaten about in the same manner. After saving him again, I re-released him and he returned once more to the same group. He has now lived with that same group for over four years and the savage beatings have not occurred again.

I was keen to find the reasons for his wounds after the first severe punishment and so watched him for hours in his interaction with others. It was clear that he did not observe the proper etiquette or rather that he had no due regard for his low status. Indeed, he tried to snatch food items from the resident male and then followed him begging for food. This could suggest that perhaps he was not very competent at finding his own food. It could also (and more likely) suggest that despite having been raised with other magpies, he was also inept at observing social rules. It is likely that the combination of those two actions, if they happened regularly, would have resulted in his demise. However, after being released the second time, he sustained no further injuries leading me to suspect that he had changed his behaviour.

His behaviour had indeed changed. To test his skills at polite behaviour, in his case subordination, I provisioned the magpie group with some food items. The problem magpie, once seen snatching food from the resident male, now stood completely still and did not once attempt to pick up any morsels, even if these fell directly at his feet, as long as the remainder of the group was present and the resident magpie was feeding. In fact, the dominant male walked over to the now submissive magpie and picked up every one of the food morsels around his feet and then flew away. Once that adult male was gone, I again provided food and, this time, the subordinate magpie eagerly collected the items and devoured them instantly.

Further evidence of the magpie's low status (and his acceptance of his status) came by observing that he did not once vocalise in the group, either in alarm calling or in carolling. The dominant male or sentinel (a second male magpie and close associate of the other male) would typically issue alarm calls or commence carolling, indicating some pecking order with respect to who can carol and in which sequence. The same restrictions do not apply in pairs. Either the female or the male can start a carolling bout. This most subordinate magpie would vocalise only distress calls and was limited to the same range of vocalisations as are juveniles. When the dominant male approached he froze and, if the dominant male continued to walk towards him, he started shaking and vocalising in distress. Intimidation encounters of this kind, and the range of responses available to the subordinate magpie, have also been described for magpies in New Zealand.[11]

Evidently, even under such unenviable circumstances, it must be of some advantage for the subordinate magpie to remain a group member. Habitat quality (if it is a prime site) is a likely explanation for staying with the group and protection from birds of prey may well be another reason. In magpie society there is also always the chance of advancement – an opportunity to move up in the social hierarchy not so much as a status symbol but as a member of the team that can be relied on. One simple way of measuring this claim is to observe what happens when that individual issues an alarm call – if that individual is inexperienced or not trusted, nobody will respond to it. Hence, if the individual makes appropriate behavioural changes, a life of relative misery (but safety) may not be forever.

One man from the Canberra region rang to let me know rather excitedly that a group of magpies was 'holding court'. He said that he had never seen anything like it and could not believe his eyes but the behaviour he was going to describe to me, he had observed once. He explained that a large group of magpies (between 10 and 20) gathered in a semicircle on the ground, all facing a single magpie at the middle of the diameter of the semicircle. The individual magpie, so he said, showed fear but did not attempt to fly away. Then magpies would start stepping forward individually one by one and peck the individual quite hard. One after the other did so and when everyone in the front row of the semicircle had had a peck at the singled-out individual they flew away, leaving the injured and defeated individual behind.

Since I had heard this fascinating story, I started watching out for such behaviour myself and, indeed, observed a very similar event in a smaller group. No doubt this may well be a rare event (I have seen it only just once in 20 years). Still, this is something that requires explanation. If, what he and I observed, is not entirely misinterpreted, this would be the very first behaviour ever described of an apparent 'rational' deliberate act in an animal.

The question of rationality in animals is anything but new and has occupied philosophers for centuries. René Descartes presumed and stated (in his *Discourse on Method, Meditations and Principles*),[16] with disastrous consequences for animals, that animals cannot think and hence humans had neither moral obligations towards nor responsibility for them. Charles Darwin,[17] 200 years later, however, more than implied the opposite. He expressed the view that he could not perceive all that much difference between humans and animals, making himself a target of ridicule and sharp criticism while people, in support of Descartes, readily and incorrectly believed well into the 20th century that animals cannot think. Thinking and rationality are not the same, of course, and the matter becomes quite complicated. There are also deeply problematic definitional problems because of the ways these terms have been used and thought about as a staple diet of philosophy for so long. Definitions of the rational versus associative processes, social behaviour versus cognition, and many more are in themselves often difficult to objectify.[18]

Whatever these interactions may mean, these events between the magpies involved no anger and no aggression, they were orderly and coherent – very much like the way judicial systems function when carrying out punishments, physical or otherwise, for past misdeeds thus involving an emotional and physical distance between an act and the response. Recently, there has been renewed interest in the subject of self-control and several publications have applied this to animals.[19] This interest is related also to an interest in cognition and decision-making and future planning.

As described in Chapter 4, we already know that the left and the right sides of the brain, while having the same nuclei, have different functions – the right side being responsible for the expression of emotions and the left side for inhibiting or modulating such emotions. Therefore, a bird's brain is already set up for suppressing emotions, hence for exercising 'self-control'. To what extent this applies to decision-making is another matter. A large-scale study by Mclean *et al.*[20] compared 36 species (birds and primates) and concluded that absolute brain size is a predictor of motor self-regulation. The larger the brain, the more likely is such self-control. This was found in great apes in particular but when birds were examined, the idea of absolute brain size did not hold at all. Birds may have as much as 90 times smaller brains than chimpanzees and yet they performed the test tasks as well as the great apes. Clearly, absolute brain size is no overall predictor of motor self-regulation across a wider range of animal taxa.[21] As explained before and elsewhere,[22] there are factors such as neuron density and connectivity, activation of certain neurons and hormones at the right time as well as the importance of lateralised functions of the brain, that cannot be ignored in any explanation.[23]

This fascinating story of the magpies holding court raises all these issues. We would need to systematically test magpies to learn how they perform on self-restraint tasks generally and what the mechanisms are in this species to enable such behaviour. The fact that it has been observed suggests very remarkable qualities and a similar complexity of social behaviour as we may also find in dog packs, some cetaceans and primates.

Cooperative behaviour

Cooperative behaviour in birds is widespread and occurs across a large variety of species and social contexts. There are birds that live communally, such as many shorebirds, ostriches,[24] and a significant number of passerines. Even nomadic and semi-nomadic species, such as budgerigars, many parrot species and some corvids, may live, feed and breed in groups or even large flocks. Some birds also engage in cooperative defence with other species.[25,26] For instance, noisy miners often actively collaborate with magpies in dealing with intruders, including birds of prey, but none of these behaviours are termed cooperative behaviour. In the literature, this term is reserved for those events and species where proof exists that not only are birds other than the parents at the nest site of a new clutch but are actively helping in feeding them. In many magpie groups this is indeed the case and much has been written about the magpie brand of cooperative behaviour.[27,28] Indeed, it was said before that, worldwide, less than 3% of songbirds are cooperative within this definition[29] but, in Australia, that number jumps to 22% or 115 identified species.[30] When one looks at the ancient lineages of Australian birds, cooperative behaviour is present in about half of all native songbird species.[31] This indicates it is a significant behaviour with a long evolutionary history on this continent.[32,33]

Cooperative breeding among birds is a subject of enduring interest because it raises the question of why any bird would volunteer to raise young that are not his or her own.[34] Dawkins argued that helping behaviour makes sense only if they have an advantage for selfish genes to continue, i.e. if helpers are siblings of previous clutches and their genes are shared by the ones they are helping to raise.[35] Increasingly, it has been shown, however, that the assumed relatedness of the helper to the nestlings or the parents is not necessarily upheld in genetic tests[36] or carries advantages as supposed.[37]

One form of cooperation extends to building nests together. Grey-crowned babblers, *Pomatostomus temporalis*, and apostlebirds, *Struthidea cinerea*, of eastern Australia cooperate in building a nest.[38] Ten or 12 babblers will combine to build half-a-dozen nests and also cooperate in feeding.[39] White-winged choughs[40,41] and superb fairy-wrens[42] are two other well-studied cooperative species.[43]

Cooperative breeding in magpies appears to be sporadic and confined to special situations and, in some cases, to special locations.[44] Usually, raising the young is undertaken by the parent magpies alone.[45] However, this is not uniformly so. In inland northern New South Wales I have seen a pair of Australian magpies having a helper at the nest and other researchers have described cooperative behaviour in Victoria in detail.[46] They have also found females depositing eggs in other females' nests (see Chapter 5). The latter may not be a sign of overt cooperation but of covert maximisation of offspring survival of the dominant hen, yet their research evidence also shows that genetic relatedness is not the criterion for helping at the nest.[46]

Chapter 6 already broached the subject of joint territorial defence. Indeed, many sedentary species actively collaborate in territorial defence, hunting, food location and in the maintenance of effective warning systems. Magpies fall into an interim position. They do not hunt together but they may forage side by side. They occasionally collaborate in locating food and in breeding, but not always. Territorial defence is one activity in which all magpies of a group, regardless of their status within the group, participate in and carry out together.

'Cooperative' behaviour has been earmarked for reproduction but cooperative behaviour in magpies is actually at its strongest and most consistent in territorial defence. Sighting a bird of prey will see all group members fly at once to the area where a sentinel has issued an alarm call and they will proceed immediately to fearlessly mob the intruder in a well-coordinated and unrelenting attack. Beak clapping and wing swishing are two additional warning signals that are constantly emitted. Once the intruder takes to the air, at least two magpies, often more, will follow and harass the raptor to a distance well outside the magpie's territory. They will accompany the bird of prey also to an altitude at which safe passage for other birds is normally guaranteed. The close flying next to a goshawk, falcon or an eagle (wedge-tailed, little or sea eagle) is not without danger for magpies. Birds of prey can 'roll over' in flight and strike with their talons sideways and upwards and magpies are 'meal size' for all of the larger Australian raptors. Therefore, to minimise this risk, the convoy of magpies will position itself simultaneously on the left and right side of the raptor's head (see Plate 11). Should the raptor try to attack one bird, the other will be able to launch an immediate counter-attack.

Ultimately, many species benefit from the magpie's cooperative vigilance. Wherever there are permanent territorial magpies, birds of prey have a more difficult time avoiding detection and they risk expulsion from the area.

Magpie territorial defence is a highly important component of their ecology. Magpies are the police of the bush (meaning open woodland) and magpie society affords some measure of protection for countless other species. It is clear from the commonly seen response by other birds, as different as noisy miners and galahs, that the alarm calls of magpies are known and heeded. Most other birds will hide when they see danger. Instead, the magpie will go out and actively, even fearlessly, pursue it.[47] Without a strong cooperative group arrangement, they would not be in the position to tackle enemies sometimes much larger in size than they are. It is the strong group cohesion and the collaboration that helps them survive.

Endnotes

[1] Kaplan 2008b
[2] Jurisevic 2003
[3] Sasaki *et al.* 2006
[4] Feduccia and Duvauchelle 2008
[5] Bekoff 2001; Pellis and Pellis 2009
[6] Kaplan 2015
[7] Ficken 1977; Auersperg *et al.* 2015
[8] Pellis 1981a
[9] Pellis 1983
[10] Pellis 1981b
[11] Veltman 1984
[12] Carrick 1972
[13] Hughes *et al.* 1996
[14] Hughes *et al.* 2003
[15] pers. observation
[16] Descartes 2004 [1637]
[17] Darwin 1904
[18] Hurley and Nudds 2006
[19] Beran 2015
[20] MacLean *et al.* 2014
[21] Kabadayi *et al.* 2016
[22] Kaplan 2018b
[23] Jelbert *et al.* 2016
[24] Bertram 1992
[25] Stacey and Koenig 1990
[26] Dugatkin 1997
[27] Mirville *et al.* 2016
[28] Pike 2016
[29] Feeney *et al.* 2013
[30] Heinsohn and Double 2004
[31] Cockburn 2006
[32] Cockburn 1998
[33] Boland and Cockburn 2002
[34] Stacey and Koenig 1990
[35] Dawkins 1976
[36] Hughes *et al.* 2003
[37] Downing *et al.* 2017
[38] Woxvold *et al.* 2006
[39] Blackmore and Heinsohn 2007
[40] Rowley 1978

41 Heinsohn and Cockburn 1994
42 van Asten *et al.* 2016
43 Dunn and Cockburn 1999
44 Hughes *et al.* 1996
45 Veltman 1989
46 Finn and Hughes 1995
47 Rowley 1972
48 https://www.youtube.com/watch?v=qoaEBb4IN4Q
49 Kaplan 2018c

10
Song production and vocal development

Songbirds are characterised by their syringeal muscles, how many pairs of these they have, and by their ability to memorise and reproduce learned vocalisations. Magpies produce some of the most complex songs and have one of the largest range of vocalisations of any songbird. As already shown, their song is not specific to the breeding season and is performed by both males and females. Magpies can project sound at very high decibels, have a large frequency range (in musical terms: 4 octaves, and in decibels from 25 to 100 dB – the latter at the noise level of a jackhammer!) and, judging by the sonagrams of magpie song, may be able to produce entirely different sounds simultaneously. Bird vocalisations, including sounds and songs, are guided by two systems working together: a primary, or central system (the song control system located in the brain) common to all songbirds, which consists of specific nuclei, and second, the physical features necessary. In songbirds these are: the main vocal apparatus, the syrinx; a secondary, or peripheral, system[1] which includes the respiratory process; and also the musculature, the trachea (windpipe), air sacs, and the larynx. Opening and closing of the beak also affects song production.[2]

Sound production

Sound production involves an exceedingly complex interaction between all of the above elements, and not all aspects are as yet fully understood. Indeed, the system becomes rather enigmatic when we deal with the degree of complexity that is typical of songbirds such as the magpie. Although the basic principles of avian sound production have been established, to gain a fuller understanding of the precise mechanisms that the magpie's exceptional abilities[3] required, we tested singing magpies in our laboratory. We wanted to establish the mechanisms of magpie sound production as well as account for its exceptional versatility.

While these results are a little technical, and details are published elsewhere,[4] it is worth summarising the findings here. Basically, the question is: how do they do it? One of the questions I was particularly interested in is how they can maintain song for so many hours. Another question concerns the enormous force and versatility of sounds and how they manage to produce noise as well as pure tones in quick succession. We measured these activities at the level of the syrinx.

The syrinx

The vocal apparatus of birds is substantially different from that of humans; yet, despite these differences, humans and birds have the most varied vocal abilities of any species (except for some marine mammals). The major difference between the mammalian and the avian vocal apparatus is in the organs involved in sound production. Humans, and other primates and mammals, rely on the larynx, an organ placed in the neck, close to the mouth. Like humans, birds have a larynx but their main vocal apparatus is called the syrinx, which consists of a collection of specialised muscles and vibrating membranes. The syrinx is hidden deep in the body, at the end of the trachea close to the forking of the bronchial branches that go to the lung on each side of the body. Therefore, a bird has two exhaling airstreams impinging on its vocal organ rather than one as in humans, and each is innervated separately.[5]

Voiced song appears to originate from vibration deep in the syrinx, mediated by the labia (lateral and medial labia) as well as by membranes (the medial tympaniform membranes).[6] These internal membranes are housed within an air sac in the pleural cavity (the inter-clavicular sac), making them sensitive to the air passing through from the lungs on expiration. The membranes are controlled by the syringeal muscles and by air pressure surrounding the membranes; it seems the elasticity and complexity of the membranes may determine the quality of sounds.

The onset and termination of vocalisation (called phonation) is usually controlled by the syringeal muscles that open or close the lumen on each side of the syrinx. The air pressure, the muscles and the internal membranes interact to produce near pure tones (single frequency similar to human whistles) and also, as in the lyrebird, to produce parallel notes, seemingly played on two instruments at once.

The syrinx varies in complexity from species to species. In the Australian kookaburra, with its loud and raucous call, the syringeal muscles are barely

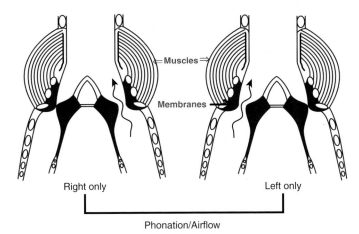

Fig. 10.1. Diagram of airflow of syrinx. The songbird syrinx is surrounded by densely packed pairs of muscles on each side, as marked, for rapid and powerful contractions and sudden expansions. Non-song birds also have a syrinx but not the muscles. The side not receiving the airflow is shut. The curved arrows show the direction of airflow.

developed. In comparison, magpies, or oscines (the true songbirds) have at least four pairs of syringeal muscles, and their syringeal muscles and internal membranes interact within this very complex syringeal system.

In most songbird species, the left and right sides of the syrinx act as independent sound sources for the production of most vocalisations, although, to a much lesser extent, both sides of the syrinx may contribute to some syllables.[7,8] Simultaneous use of both left and right sides, at least in the species that have been studied, produces syllables that are 'two voiced', i.e. are not harmonically related and of different amplitude modulation.

A diagrammatic view of a tranchobronchial syrinx (Fig. 10.1) shows the direction of airflow before it hits the membrane that produces the sound. Indeed, the fact that crossing the membrane is bifocal theoretically would allow the bird to stimulate each area separately.

I have autopsied ill-fated magpies (accident victims) for their syrinx and found that the syringeal muscles of the Australian magpie are very noticeable and appear to have a pronounced asymmetry, with the left-side muscles larger than the right. Such asymmetry is consistent with studies on other species, both in terms of actual arrangement on each side as well as in terms of their varying strength and thus function (Fig. 10.2).

During singing, produced during expiration, magpies can switch song production between the two sides of the syrinx, i.e. sound can be produced first by one side of the syrinx and then the other. There is evidence that, occasionally, the bird can use both simultaneously to produce harmonically unrelated sounds. These data suggest complexity of magpie song involving the rapid execution of

Fig. 10.2. Syrinx of an adult magpie. The trachea (tubular) merges seamlessly into the bronchial half rings surrounding the syrinx before they divide into the left and right bronchi leading to the lungs. During exhalation, air proceeds from the lungs to the bronchi, across the membranes of the syrinx and via the trachea across the larynx into the mouth. The syringeal muscles are located just under the sternum (beginning of the rib cage).

complex syringeal motor patterns that are independent from, but coordinated with, those of respiratory muscles.

Further, by quickly opening both sides of its syrinx for a bilateral mini-breath, the bird minimises the duration of the silent period required between syllables. Some magpie vocalisations have prominent nonlinear characteristics including broadband vocalisations (chaos) and subharmonics. Since nonlinear phenomena like subharmonics arise from the passive physical properties of the sound source (the labia can oscillate in different modes), they may provide a cost-effective way of increasing vocal complexity without requiring complex neural circuits for motor control of the vocal organ.

These mechanisms also partially explain why magpies can sing with little interruption and seemingly little energy expenditure for many hours at a time. When a cross-section of all parameters is examined, one can see how ingenious the system for sound production is (Fig. 10.3).

The magpies demonstrated bilaterally independent control of both the timing of phonation and the frequency composition of sounds as these were generated on each side of their syrinx.

Although the majority of syllables included contributions from both sides of the syrinx, and produced the same fundamental frequency (f0), there were other

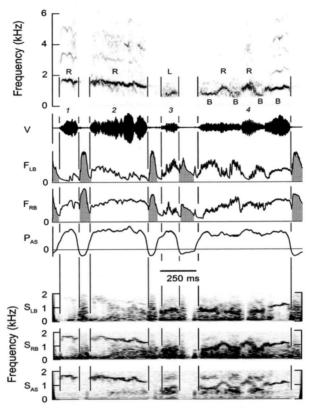

Fig. 10.3. Sound parameters. The entire segment presented here is a cross-section of all sound parameters captured within a span of a mere 1.5 second window, substantially enlarged, recording at the level of the syrinx in eight simultaneous measures, i.e. to be read vertically. The top part of the figure shows four syllables, separated by minibreaths. Next panel below: Vocalisation (V) time waveform with syllables numbered; third down: Rate of airflow through left (F_{LB}) and right (F_{RB}) sides of syrinx. Horizontal line zero airflow and inspiratory flow is accompanied by negative air sac pressure and is indicated in grey shading. Next line down: Sub-syringeal pressure in the thoracic air sac (P_{AS}). The last three panels (S_{LB}, S_{RB} and S_{AS}) show sound in left bronchus, right bronchus and thoracic airsac. (For full details see Suthers et al.)[4]

times when the frequency from each side was different. This is the so-called two-voice vocalisation. This ability has also been shown to exist in the brown thrasher, *Toxostoma rufum*, and grey catbird, *Dumetella carolinensis*, both of the family Mimidae and among the most vocally versatile passerines of the Northern Hemisphere[9] and it is extremely rare, as far as we know to date.

A striking feature of many magpie vocalisations is the presence of prominent amplitude modulation (AM) at high modulation rates exceeding 100 Hz. Some of this amplitude modulation is a beat signal produced by sounds emanating from each side of the syrinx at slightly different frequencies.

In contrast to most other songbirds, in magpies the higher fundamental frequency during two-voice syllables was usually generated on the left side of the syrinx. Amplitude modulation was produced by interactions between different frequencies that may originate either on opposite sides of the syrinx or on the same side. Hence the mechanisms are nothing short of ingenious – creating in the syrinx a musical instrument of great flexibility because each side can act independently, make switch overs and has the muscle power to convert these vibrations into remarkably far-reaching signals and beautiful song.

The trachea

The trachea filters sound, so its length is important for sound production. In magpies, not surprisingly, it is much longer than, for instance, in a canary or zebra finch. Indeed, from the syrinx to the entrance of the trachea the mean measure of the tract was 90 mm ± 5 mm, longer than the entire body of a zebra finch. There is an inverse relationship between the length of the trachea and the ability to produce fundamental frequencies: the shorter the trachea, the higher the sound and the smaller the range of fundamental frequencies produced; but the longer the trachea, the lower the sound and the greater the range of fundamental frequencies that can be produced, as is the case in magpies. The same inverse principle applies to musical instruments – the long body of a cello versus the short body of a violin; the latter produces higher notes, the cello can also produce very low notes.

Musculature

Any keen bird observer would be familiar with the sight of magpies positioning themselves and adopting a distinct body posture before carolling, a loud acoustic signal typically in the range of 6–8 kHz. In such vocalisations, the bird widens its chest, by splaying the breast feathers, tilts the head backwards and reclines its entire body backwards a little. This suggests that muscular activity for the production of the actual signal, and for amplitude in particular, comes from a variety of sources, not just the syringeal muscles. These include the chest muscles, neck muscles and even back muscles. Therefore, producing a specific sequence of sound at a certain frequency range and amplitude requires a remarkable coordination of large parts of the bird's musculature in addition to the sound-producing organs (Fig. 10.4).

Song control system

After more than three decades of research into the brains of songbirds, we now have a general 'map' of the brain indicating which parts of the brain are involved in song production.[10,11] In birds, song learning and production are controlled by a network of forebrain nuclei. They include what is called 'the high vocal centre'

Fig. 10.4. A, Side view and B, frontal view of magpie carolling. Carolling requires a change of posture, throwing back the head and engaging musculature for producing sounds that can carry far. Note the extended feathers on throat and belly.

(HVC), the so-called robust nucleus of the archistriatum (RA), Area X, and the lateral magnocellular nuclei of the anterior neostriatum (lMAN) (see Fig. 10.5).[12,13]

The map is a very good guide but there are a tremendous number of variations possible. Having established that certain nuclei in the brain are vital for song production, researchers turned their attention to the connections between these various nuclei, the role of hormones (particularly testosterone), and environmental factors such as daylight variation.[14]

Much of the research has been carried out on zebra finches and canaries, *Serinus canaria domestica*.[15,16] Zebra finches and canaries are sexually dimorphic, particularly in song and partly also in plumage. In zebra finches only the male sings in order to attract a female and in canaries the female does sing but only the male has the elaborate breeding song. These differences are reflected in the size of the nuclei. For instance, male zebra finches performing breeding songs have a well-defined Area X but that of the non-singing females is much smaller.[17]

Moreover, the HVC and neuronal types in HVC of the female zebra finch differ from those of the male.[18] Structural differences in brain nuclei have been shown for functional differences in canaries[17,19] and European starlings.[20]

However, relatively little attention has been paid to the structure of the song control system in those bird species in which both the male and female sing and behavioural dimorphism in song may be weak.[21] In magpies, both males and females sing (not using song for breeding or specifically for courtship). For many

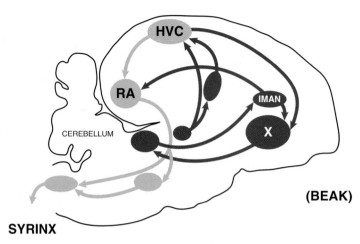

Fig. 10.5. Vertical cut of the brain showing the nuclei involved in song production. In order to produce vocalisations, different parts of the brain are activated. The high vocal centre processes sounds and passes these on, with several feedback loops, to the motor system to enable the syrinx to produce the sounds. RA (robust nucleus of the archistriatum), HVC (high vocal centre), X (Area X), lMAN (lateral magnocellular nuclei of anterior neostriatum).

years, such female song was regarded as a mere oddity of some Southern Hemisphere birds but finally, it seems, it is recognised that female song cannot be ignored any longer and may be of evolutionary significance.[22] Of course, we knew this with absolute certainty by 2001, when we were able to confirm in a paper that magpie males and females do not just sound alike in song but, in fact, the female has the same song control system that has been shown to exist in seasonally singing males. Research in our laboratory also compared the nuclei of magpie male and female, as well as juvenile and adult brains.[3] The research, based on a sample of nine magpies, identified the song control nuclei, including HVC, Area X, RA, and lMAN, and found the same network in the forebrain of canary and zebra finch males. However, unlike canary and zebra finch males, these nuclei of the forebrain were found in both male and female magpies and they were as well developed in females as in males. The HVC of both males and females, as well as of juveniles and adults, has a similar structure. All of these nuclei were also well developed in juveniles aged two to three months post-fledging, with females showing earlier maturation in some nuclei than males. This is an impressive instalment in the brain – it is puzzling why it is there at all since it is not used for breeding or territorial defence.[23] In other words, singing in magpies does not seem to bestow a breeding advantage to the male or the female.[24]

Since magpies sing all year round and males and females sing alike, it is perhaps not surprising that the nuclei in the adult forebrain show full development in the first year of life, with little to no sexual difference between male and female adults.[3]

Many theories on vocal development draw on a close association between behaviour and neuroscience[25] since behavioural development is dependent on physiological and neural processes. Sound production in songbirds also depends on the vocal apparatus and its development: in some species full song needs to be learned first (if they have a limited time to learn) and is first produced only once the syringeal muscles have fully developed.[4] In lifelong learners, as magpies are, one can expect new sounds and syllables also to appear in adulthood. These are some of the prerequisites for song and some of these grow, alter or are subject to transitional stages during development.[26]

There are many recorded observations about the vocal repertoire of Australian magpies.[27,28,29,30] However, developmental studies have been more or less entirely absent and the first detailed research papers were published in 2018.[26]

The other issue is of some significance. Not every vocalisation in birds is song. Songbirds and non-songbirds make many vocalisations other than song that form an important part of the bird's total repertoire and communication. These vocalisations have been relatively neglected by neuroscience in recent decades, possibly because they are often rather stereotyped and appear to hold little promise in terms of broader questions on vocal learning, let alone cognition. There have been different reasons for this neglect, with the exception of referential signals, as will be addressed in more detail in the coming chapter.

Song development and learning

Song development in an Australian magpie is related to the development of the syringeal muscles and also to the brain itself. We found that the song nuclei of female juveniles were of adult size but this was not the case for juvenile males. Full song is produced only once the syringeal muscles have fully grown.

Audible vocalisation of magpies begins about a few days post hatching. The sounds nestlings make are very low in amplitude and of high frequency and they consist of single sounds (Fig. 10.6). By week two, the amplitude has increased substantially and first formants are beginning to appear. By this time, magpie nestlings begin to make sustained and clearly audible begging sounds. By the end of the third week, magpie nestlings begin to practise song and they do so independently of the adults.

Song practice is conducted only when the parents or any adult magpies are at least 15–20 m away from the nest, presumably just outside audible range of the adults.[31] This may be due to social reasons – because magpies are territorial, their song has to be distinguished from that of their neighbours. Although song may play no role in securing a partner, we do not know whether it has a role in territorial acquisition or defence. (We do know, for example, that the major function of female song in the superb fairy-wren, *Malurus cyaneus*, is, in fact,

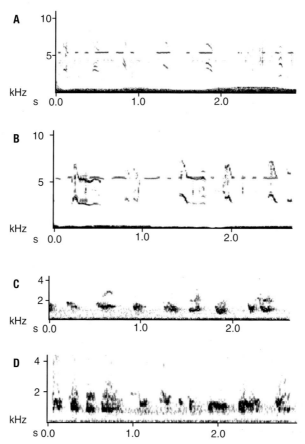

Fig. 10.6. These four sonagrams document the development of song. A, A one-week-old nestling, high pitched, unmodulated and barely audible. B, A week later modulations have begun to appear and there are the first signs of harmonics at the level of typical adult song (around 2 kHz). C, The third week brings a major transformation as a consequence of growth: the voice has moved from the high (5 kHz) to a much lower register of 1.5–2 kHz, because of the extra length of the neck and syrinx, and strengthening of the syringeal muscles. Modulation is still limited. D, At least five weeks of age. Modulations are clearly visible and sounds may involve two harmonics but there is as yet no fluidity. Each element is still staccato and no matter how hard the bird tries, it cannot produce a faster rate of sound.[19]

territorial defence.)[24] Also, individuality may need to be asserted if a bird should gain a territory and have breeding status. Indeed, magpie parents suppress vocalisations of the young in their presence and allow their offspring only to use begging and distress signals.[32] Even the begging calls are quite limited as long as the offspring are still in the nest, presumably because, as was said before, magpies feed their young equally, no matter how much begging is used by a nestling.

There is also some distinction made between simple calls (brief calls such as in distress, alarm, mobbing, contentment) and complex song and it has been argued that calls express emotional states, are not necessarily designed for communication and thus do not require learning. There has been little direct experimentation to test whether calls are learnt or not but some field observations might lead us to suggest that at least some calls are learnt – those that are used in communication.

It is possible to hypothesise that distress and begging calls in magpies are not learnt. This minimal sound repertoire may well be innate but even these calls have some communicative function. Distress calls, while having some properties that are shared among many species, have their own signature for a species[33] and a distress call by a magpie is clearly recognisable as that of a magpie and not of another avian species. Moreover, each of the simple calls of juvenile magpies that I have studied so far has slightly different sound patterns and this suggests that adult birds might recognise the identity of the juvenile caller merely by the vocalisation, and, by some structural changes in the vocalisation, may be able to assess the urgency of the call.[34] Alarm calls, as I shall discuss in Chapter 11 in more detail, may well be an example of calls that are learnt. Alarm calls may also distinguish between the type of danger to be flagged to conspecifics ('I have seen an eagle' versus 'there is a lace monitor') and this may be a complex signal.[35]

Song learning

From studies of song production in different bird species, we know that birds may acquire songs at an early age, that there may be an age limit to their learning ability, and they may need their own, and sometimes very specific, tutors. In some birds, as in zebra finches, there is just a small window of developmental time to acquire song and the most effective learners are those that have a tutor of social importance to them.[36] We know that early learning can be quite important – zebra finch nestlings deprived of song learning opportunities were found to develop very impoverished song in the following season.[37]

Young songbirds may learn a vocalisation after very limited exposure. Nightingales, for instance, learn an entire song by hearing this song as seldom as 10 times during their sensitive period for song learning.[38] Magpie juveniles appear to have a similar capability of learning new song types after only brief exposure. I have collected examples of mimicry in juvenile magpies for which evidence exists that exposure to the sounds had been very brief.[39] In one case it was possible to show that a juvenile magpie was capable of an accurate rendition of kookaburra calls ('laughing') after an exposure period of just minutes.[39] The period for learning new songs may vary substantially between species.

In magpies, plasticity (the ability to learn new songs) remains high throughout the first year of life at least. Hand-raised magpies were able to learn new sounds and new (human) words and extended their range of vocalisations continuously up to the age of 12 months.[40] Plasticity in magpies after this age has not been tested, largely because birds have not necessarily been kept for more than a year.

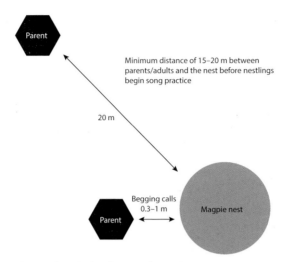

Fig. 10.7. Song practice by nestlings in the absence of parents.

However, there is every indication that the magpies may well be lifelong and open-ended learners. Some indirect evidence for this may be obtained from data on song sharing. Song sharing[41] with territorial neighbours presupposes the ability to respond to, and integrate into one's own song, new vocal requirements in a new territory in adulthood. As most territories are established only when adult birds have been able to claim a territory of their own and most juveniles eventually exchange natal territories for new ones, song sharing with neighbours may have to be an acquired adult skill.

Research has shown that magpies share as much as a quarter of their song with neighbours which indicates that the birds need to have the ability to learn these while already at an adult stage. This tuning in to the vocal landscape gives each territory its particular stamp of identification. The new territories may be merely a few kilometres away or, as recent research has found, even further away from the natal site.[42] It leads to geographic variations in vocalisations by the same species, as has been shown in magpies,[43] and presupposes the ability to respond to, and integrate into one's own song, new vocal requirements in a new territory in adulthood.

Environmental criteria are important in song learning. The fact that song is learnt during the stages of life when nutrition is critical (nestling and fledgling stages) may mean that the physical condition of the bird influences its ability to learn about song.[44]

Song learning is also influenced by many social factors.[45] In magpies, interaction with a parent or adult group may make vocal copying possible but all indications are that copying from parents is suppressed and song practice has to take place when parents are well away from the nest (Fig. 10.7).

Thus, unlike many songbirds, magpie youngsters do not learn directly from their parents and, for whatever inherent reasons, nestlings will not vocalise and practise until parents are well outside the visual field. In extended nest research I found that this is a consistent feature of all nestlings, certainly in groups that have no helpers, i.e. the adults disappear from the nest site to find food.[45] The nestling song practice is improvised and self-guided. However, since 2008 we also know that songbirds can learn by listening. They need to hear a vocalisation and a specific set of neurons, called mirror neurons, fire up as if the bird were singing the tune itself.[46,47] Interestingly, an image I have presented often because it is the only visual exemplification of the body's involvement is a juvenile listening to its mother carol. As said before, carolling is not permitted to be voiced by juveniles, but the image shows that the juvenile's body is showing all the signs of active carolling performance even without producing a single sound (Fig. 10.8).

Of course, nestlings will still be able to hear some song in the distance. The social interactions that occur between neighbouring birds might be important for this mutual learning of song. Even just seeing the singing neighbour may be important too.

It has been suggested that ecological factors may have a large bearing on what is learnt and then actually produced.[48] Many species know more songs than they

Fig. 10.8. Listening juvenile ready for carolling continues intently listening to the adult female carolling in the background but not singing itself.

172 | Australian Magpie

AGE (weeks)	1st month				2nd month				3rd month				4th month				5th month				6th month				7th month+	
	5	6	7	8	9	10	11	12	13	14	15	16	17	18	19	20	21	22	23	24	25	26	27	28	29	30
Vocal repertoire																										
Distress	♦	♦	♦	♦	♦	♦	♦	♦	♦	♦	♦	♦	♦	♦	♦	♦	♦	♦	♦	♦	♦	♦	♦	♦	♦	♦
Begging	♦	♦	♦	♦	♦	♦	♦	♦	♦	♦	♦	♦	♦	·	·	-	-	-	-	-	-	-	-	-	-	-
Warble	·	·	·	·	·	·	◆	◆	◆	◆	◆	◆	♦	♦	♦	♦	♦	♦	♦	♦	♦	♦	♦	♦	♦	♦
Threat/Disapproval	-	◆	◆	◆	◆	◆	◆	◆	♦	♦	♦	♦	♦	♦	♦	♦	♦	♦	♦	♦	♦	♦	♦	♦	♦	♦
Mimicry	-	-	-	-	-	-	-	-	·	·	◆	◆	◆	◆	♦	♦	♦	♦	♦	♦	♦	♦	♦	♦	♦	♦
Alarm calling	-	-	-	-	-	-	-	-	·	·	·	·	·	·	·	◆	◆	◆	◆	◆	◆	◆	◆	◆	♦	♦
Carolling	-	-	-	-	-	-	-	-	-	-	-	-	-	-	-	-	-	-	-	-	-	-	-	-	-	-

Fig. 10.9. Vocal development post-fledging and development of other behaviour. (-) never observed; (·) rarely observed or tentative; (◆) increased activity but not always successful and/or not fully developed; (♦) fully developed/expressed.

use and some others first learn a number of songs but then shed most of them before they arrive at the breeding site. Some birds may be left with just one song, as is the case in white-crowned sparrows, *Zonotrichia leucophrys*.[49] In the magpie, it may well be the other way around. They (male and female) start off modestly and gradually build up a substantial repertoire of songs. By the time they acquire their own territory, they are often at least five years old and they may well face new neighbours with new repertoires that they then partly incorporate. Since magpies are sometimes forced into serial occupancy of different territories, one may surmise that some vocal adaptation to the new environment would be required in each case.

A final point on the development of vocalisations in magpies. Not all sound groups develop at the same time. There is even a developmental and a social timetable for their expression (Fig. 10.9).

Mimicry and learning

There is ample evidence that many songbirds mimic and some of them, such as the migratory marsh warbler, *Acrocephalus palustris*,[50] and the European starling, *Sturnus vulgaris*[51,52,53] may well hold records for producing the largest number of identified mimicked sounds. Some 113 species are mimicked by male marsh warblers and as many as 11 orders of birds by starlings, including also the vocalisations of mammals and the sounds of inanimate objects.[51] Alec Chisholm was the first to list 56 Australian avian species as having some ability in mimicry.[54] Since then about half of the ones that he named have been verified.

Vocal mimicry is so far known to occur only in birds, some cetaceans and seals, as well as in humans.[55] Indeed, all human infant learning of speech sounds initially consists of mimicked sounds, getting the intonation right and going through an extended babbling stage, then formulating sounds that begin to sound like small words in the language environment into which it is born. The crucial step (actually more than one) is the ability of the human language-learning infant to use the word learned in a correct context, i.e. give it meaning and a framework created for communication.

According to statements made some 50 years ago, vocal mimicry was defined as the imitation by birds of sounds outside their specifically characteristic vocalisations.[56] Armstrong[57] similarly argued that mimicry is 'the copy of sounds other than those of the bird's own species'. This basic definition has not changed but what we know about its potential functions has changed. Armstrong was right to include the comment 'other than its own species' for two main reasons. First, mimicry provides clear evidence of the ability of vocal learning because vocalisations of other species are simply not part of an inherited template. Second, mimicry may well be regarded as a paradox because the entire system of bird vocalisation is built on species specificity.[51]

Nonetheless, mimicry is extremely widespread among Australian birds although not all that well documented to date.[54] The best-known Australian examples of mimicry in the wild are for Australian magpies and for lyrebirds.[58,59,60,61] The magpie is an excellent mimic, equally versatile in producing mimicry as often heard in the male lyrebird's vocal displays.[61,62] In contact with humans, even if remaining free, they can also mimic human speech. We know that parrots and budgerigars are excellent mimics in captivity but this may be an artefact, although, for the African grey parrot, examples of some mimicry in the wild have been found.[63]

My first data of magpie mimicry were derived from recordings other people had made all over Australia and, thus are not specific geographically. Mimicry may therefore represent a skill and practice found in possibly all of the magpie subspecies (I have, however, no samples from the Northern Territory and Tasmania). Magpie mimicry samples that were submitted included cats, dogs, horses, humans, other birds and even potential predators (such as the barking owl). My sample of 20 separate records from Western Australia, South Australia, Victoria, Canberra, New South Wales and Queensland found that all mimicked sequences were linked to species that occurred at the time within the territorial boundaries of the mimicking magpie. This is not as self-evident as it sounds. After all, there are always transient vocalisers in magpie territories, such as visiting or migrating birds, nomadic birds, cats, dogs, horses or humans passing through.

On the basis of this Australia-wide information received, I started my own investigations. As Fig. 10.10 shows, one can assess the size of the territory, where the boundaries are, observe and record other species being present in the territory on a permanent, regular, seasonal or just occasional basis and then record any

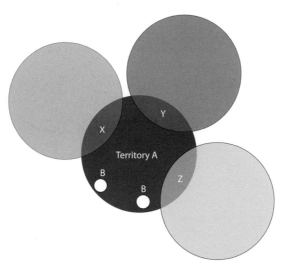

Fig. 10.10. Sound context of mimicry. The dark central circle indicates the territory of the magpies under investigation. At the southern end (marked B) are two other magpie territories (white circles) where skirmishes occasionally occurred. The surrounding grey circles indicate other territories. Some species permanently share the territory with the magpies (e.g. horses, sheep, dogs, humans and other avian species). X, regular visitors (such as migratory birds arriving and staying for the entire breeding season); Y, regular visitors but not on a weekly basis or for any length of time; Z, occasional visitors or transients.

mimicry to ascertain whether any species (birds and other animals, also including humans) within or near the territory were being mimicked.

While the results of my research were published elsewhere,[64] suffice it to say here that magpies only copied sounds made within their territory and disregarded categories of rare or occasional visitors. It is as if the bird, in addition to a clear geographical map of its territory, also constructs its own vocal map. Unlike the lyrebirds, magpies do not usually appear to mimic inanimate objects, at least not as often as do lyrebirds (such as chainsaws or car horns). Magpies do not string their mimicked sequence along a fixed sequence as lyrebirds do but intersperse their own song with snippets in any manner possible. More importantly, magpie mimicry seems to be largely related to territoriality while, in lyrebirds, it is all about sound effects. I have found no evidence that mimicry is confined to specific seasons, as has previously been suggested.[65]

Many years ago I analysed the quality of mimicry in magpies and lyrebirds.[39] The Australian magpie and the lyrebird lend themselves to comparison. Both have an ancient lineage among the Australo-Papuan centred corvida.[66,67] Both are territorial and ground feeders, and both are capable of producing loud and musical notes that are widely audible. Their frequency range is similar and their skills in and extent of using mimicry are similar, although the lyrebird has only three syringeal muscles, rather than the oscines' typical minimum of four pairs.[61]

There are several other major differences between lyrebird and magpie renditions of mimicry that were identified. Lyrebirds sounded elegant, loud, melodious and utterly convincing to the human ear, including also to some local species such as kookaburras that actually often reply to passages of their song being mimicked. As a performer, and a dazzler, the lyrebird cannot be beaten, not even by the magpie. And in lyrebirds, the song has a function, it is the male's breeding song and is meant to impress a female to mate with him. The performances have to be racy and impress, they do not need to be all too precise.

By contrast, in terms of tempo, rhythm, amplitude or modulation, magpies were more accurate in their renditions than lyrebirds, this is evident by scrutiny of sonograms.[68] But despite this, the magpie sometimes does not sound very convincing. For instance, I had the rare opportunity to record a kookaburra duet and, the next day, found that the magpie was suddenly mimicking kookaburras and realised that this was the mimicked rendition of this very duet the magpie had heard the day before. It is evident from the recording (and from listening to the magpie sing) that the bird struggled a little to mimic the staccato of the kookaburra and it all sounds a little amateurish (a 'C+' at best if the magpie's performance were rated against the lyrebird at the top of the class for all performances) (Fig. 10.11).

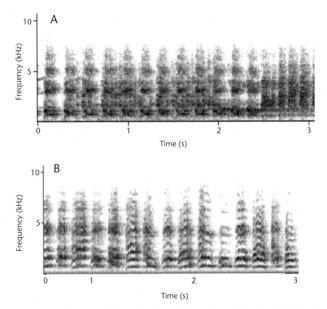

Fig. 10.11. A, Recording of kookaburra duet and B, magpie mimicry of the duet. Interestingly, the magpie attempted to copy the duet rather than the individual bird. Note that the magpie copied the exact number of staccato calls in exactly the same amount of time. Hence the magpie's rendition is completely faithful to the rhythm. The sound intensity is also matched precisely. It also tries to sound husky (introducing more noise into the call than magpies would have in their own song), placing the energy of the call at the same frequency levels as the kookaburras (darker lines at ~1 and 5 kHz respectively. Still, the magpie's sequence lacks the extensive 'noisy' components of the kookaburra calls.

Magpies do a very good rendition of a horse neighing, a dog barking and of a barking owl's signature 'bark', presumably because these are nowhere near as energy demanding as a kookaburra duet!

The question is, does that make mimicry in magpies mindless? Do parrots parrot and apes ape? A number of functions for mimicry have been identified in other songbirds[69] but not in the Australian magpie.

Mimicry and cognition

For those who are interested in cognition in birds, it is important to note there are now more and more examples being found that some of the mimicry that birds acquire may not be 'mindless' at all, either because such mimicry may have important functions (even be deceptive) or because they have been applied creatively in new situations. There are examples that magpie mimicry may not just be rendered as mimicked sounds but applied in new contexts which suggest a translation process from sound to meaning, as happens in infants. We have very few examples of being able to validate whether a magpie somehow understood how to use a word or sound in an appropriate and new context. Indeed, worldwide there are very few studies of songbirds that have been tested for their comprehension of the human words that they had incorporated in their own vocal repertoire. Irene Pepperberg[70] devoted a considerable number of years to showing that certain species can, just as human infants, begin to use sounds in ways that have meaning in specific contexts. Her African grey parrot, Alex, for instance, answered questions correctly in English and also seemed to ask questions in the right context. My pet galah I had inherited when he was already over 70 (human) years old took to complaining about my absences at night by asking, the minute I entered the house, 'Where have you been?' and then turned away in feigned anger and would only reconcile if I spoke to him directly. In free-ranging magpies this is of course difficult to establish but quite a few of them befriend people and may pick up words they then incorporate. There is one that I have cited quite often:[71] a magpie had learned to use the name of a dog on a property in New South Wales. The owners also had a cat that had tried everything to get rid of the magpie. When the cat approached, the magpie did not fly away but called out the name of the dog; the dog came running and chased the cat away. Calling the dog was not mimicry any longer but a most effective way to use the dog's name. Nick Fox related a story to me recently of a hand-raised magpie that lived on a farm near the shearers' quarters where falconry meetings were held some years ago. While the harriers were shackled and had a cover over their eyes, one bird pumped its wings near the magpie creating strong gusts nearly knocking the magpie off its perch – the magpie looked up and shouted 'Piss off!' clearly and loudly to the amusement of all present. There are other small incidents that could be cited. It seems to me that human infants start using words when in adult company, and when they happen to hit the right word for the right context, it might be greeted with laughter, cuddles or social encouragement and such shaping of responses eventually leads to an expansion of their vocabulary and, more importantly, to a use of that vocabulary in novel but correct contexts. In the shearers' shed, the magpie's expletive brought the house down. In pet birds, especially cockatoos, replies that please will likely also be shaped to be repeated and, increasingly, might then appear in the right context.

Endnotes

1. Suthers 2001
2. Wild 1994
3. Deng *et al.* 2001
4. Suthers *et al.* 2011
5. Nottebohm 1972
6. Goller and Larsen 1997
7. Nowicki and Capranica 1986
8. Nowicki 1987
9. Suthers *et al.* 1996
10. Nottebohm 1970, 1971, 1972, 1975, 1980
11. Konishi 1965
12. Brenowitz *et al.* 1997
13. Wild 1997
14. Tramontin *et al.* 2000
15. Arnold *et al.* 1976
16. MacDougall-Shackleton *et al.* 1999
17. Konishi and Akutagawa 1985
18. Fortune and Margoliash 1995
19. Ball *et al.* 1994
20. Bernard *et al.* 1993
21. Kroodsma *et al.* 1996
22. Odom *et al.* 2013
23. Brown and Veltman 1987
24. Cooney and Cockburn 1995
25. Mooney 2009
26. Kaplan 2018a,c
27. Brown *et al.* 1988
28. Brown and Farabaugh 1990, 1991
29. Farabough *et al.* 1988
30. Sanderson and Crouch 1993
31. Cooney and Cockburn 1995
32. Jurisevic and Sanderson 1998a
33. Jurisevic 1999
34. Kaplan 2003
35. Weary and Krebs 1987
36. Nottebohm 1971
37. Jones *et al.* 1996
38. Hultsch and Todt 1989
39. Kaplan 2000
40. Farabaugh *et al.* 1988
41. Baker *et al.* 2001
42. Brown and Farabaugh 1990
43. Nowicki *et al.* 1999
44. Farabaugh *et al.* 1994
45. Kaplan 2005
46. Prather *et al.* 2008
47. Kaplan 2015
48. Nelson 1999
49. Derryberry 2009
50. Dowsett-Lemaire 1979
51. Hausberger *et al.* 1991
52. Hindmarsh 1984
53. West and King 1990
54. Chisholm 1948
55. Moore 1996
56. Thorpe 1964
57. Armstrong 1963
58. Robinson 1975
59. Thoburn 1978
60. Sanderson and Crouch 1993
61. Robinson and Curtis 1996
62. Kaplan 2000, 2003
63. Cruickshank *et al.* 1993
64. Kaplan 2004, 2015
65. Collins 1983
66. Schodde and Mason 1999
67. Sibley and Ahlquist 1985
68. Kaplan 1996, 1998, 2000
69. Kelley and Healy 2011
70. Pepperberg 2009
71. Kaplan 2016

11

Communication

To some extent, each chapter has discussed acts of communication, be this in terms of revealing food sources, in territorial defence and in everyday life. Birds living in socially complex groups, such as ravens, some parrots, honeyeaters, superb fairy-wrens and magpies, may develop very complex communication systems compared to other species in their weight class and some may be comparable to those of some mammals (e.g. great apes or wild dogs). Most of the more vocal behaviours are now known to be acquired through learning, and some may depend on higher cognition.[1] Mastery of communication skills, including those of vocalisation, may play a very important role in the degree of success a bird has in finding a partner, breeding and holding on to a territory.

The view has developed that animals living in close affiliative associations acquire what is termed 'social intelligence'. This is precisely because of their need to maintain close, affiliative and cooperative contacts that are thought to require more extended communication skills than those living in loose flocks.

In a recent set of interesting experiments, Australian magpies were set to solve colour-associated foraging tasks and while they did, there were significant differences. Individuals from larger groups were more likely to attempt to solve the task. Additionally, older individuals were more successful, suggesting that both social and individual differences contribute to associative learning ability.[2] Of late, social species have not only been linked with cognition but with longevity, an interesting point since so many Australian native birds, including magpies, are long-lived.[3]

Complexity in communication has also been linked to group size, closeness of affiliation and cognition. How spurious or real these links really are needs to be tested in each individual species. It seems though, that the Australian magpie, with its vast net of communication options and song may have developed many special cognitive and communicative abilities. Perhaps in their case these may well be essential in order to secure their position and allow them to create and maintain territories and thus prime properties. They repay by allowing their land to be used by other species and, perhaps inadvertently, defending other species again predator incursions.

Often when one thinks of a songbird, one thinks of its song. Not all vocalisations in birds are song. Songbirds and non-songbirds alike make many others that form an important and integral part of the bird's communication, indeed, on a daily basis they carry substantial weight. These are typically short vocalisations, chief of which are its alarm calls. Excepting a few notable analyses short calls have been relatively neglected, possibly because they are often rather stereotyped or appear to hold little promise in terms of broader questions on vocal learning. However, within this group, alarm calls have been singled out and a vast literature has arisen within this category alone, be this for an interest in acoustics, effect, communication or even cognition.

Magpies have many different classes of vocalisations in their total vocal repertoire. Even without studying these in detail, one might well and rightly suspect that repertoire size of different call and song categories may suggest a potential for the utilisation of very different message types.

As Fig. 11.1 shows, the classes of vocalisations also may have different origins. We can be fairly sure that emotional expressions, some basic alarm and distress calls are not learned but part of the inherent template of magpies and other birds, songbird or not.

But for all other vocal categories, several qualifications might be required. In sexually dimorphic males (in species in which only the male sings) song is learned by juveniles and often has to be acquired in a relatively short time window (called the sensitive period), as in zebra finches for instance.[4] Learning ability (referred to as plasticity) may decline relatively early in development once the song(s) that need to be learned have been perfected. However, in other species, learning ability may continue over a lifetime[5] and add to the bird's overall repertoire.[6] Quite a separate category are species that retain plasticity throughout life as some songbirds do, including the Australian magpie.

A further complication lies both in number and type of song, its function and even male competition or female choice that determines how much can be learned by tutoring or by a self-styled improvisation. The magpie is an improviser and can add or subtract from its song as young birds are not taught, as already mentioned.

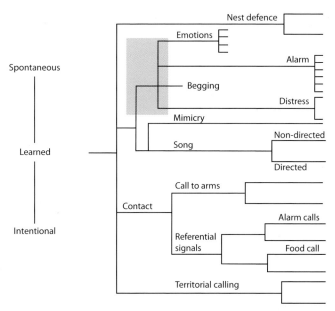

Fig. 11.1. Categories of vocalisations in Australian magpies. Magpies may express a wide range of emotions and information via vocal means. Note that alarm calls alone can be subdivided into six major categories (27 different alarm calls in total have been identified that fall largely into these six categories). The shaded area refers to call types that juveniles may also express.

The category that was most contentious, is now well established. It is called intentional communication, i.e. expressing vocally an intention to inform others about something and something very specific. For both sender and recipient it means that the sender needs to have someone in mind and for the recipient to understand that the message may contain information of importance and may even require anticipatory thinking. Most often, this is studied in referential signals (discussed below).

On the surface, communication sounds quite simple. Communication, as distinct from vocalising, somehow presumes that one bird sends a message to another and that the information conveyed is received, understood and, if need be, acted upon (Fig. 11.2). It is not quite that simple. Some 'messages' are not vocal, especially close-up communication, others are long distance and distorted. Some are heard and not acted upon.

In some bird species communication through sound is equally important or superior to visual information.[8] This need not involve sophisticated vocalisation but may require excellent hearing. It was mentioned that the magpie, able to locate underground grubs by sound alone (as explained in Chapter 5), has obviously excellent hearing in detecting sounds of low intensity. Magpies also have very strong and high amplitude calls that may carry far over a large terrain.

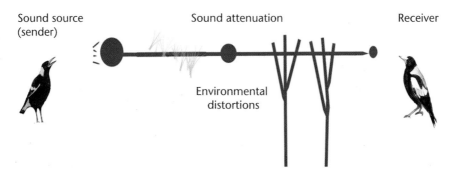

Fig. 11.2. Sender and receiver. The magpie on the left is sending a message. The large circle represents the power at which it was sent. That power can be diminished in various ways. Distance alone makes the amplitude of the sound diminish (called attenuation). Other factors can cause further attenuation or even distortion (grey squiggles). These may be invisible obstructions, such as sounds created by wind, updrafts, rain and fog, and a range of visible obstructions such as trees and shrubs, refracting the sound. The recipient of the sound may receive a signal that is much diminished (smaller circle), possibly distorted or interrupted (cf Kaplan[7]).

Regardless of the relative importance of one or the other modality, generally, no single aspect of the communication system functions entirely in isolation. At close range, communication can occur without any vocal signals. Thus, before coming back to any long-range vocal signals, it is worth exploring these non-vocal signals. These are often overlooked in discussions on communication but may have important regulatory functions in a group.

Body posture

Birds, like mammals, have a wide range of body postures available to them for communicating a particular message, and magpies certainly use a great variety of these. Magpies express agonistic behaviour by arching of the neck, extending the wings outwards, and stretching the head forward before engaging in any specific agonistic locomotion or action. Some of the body postures issue a warning not to approach and readiness to either run or fly in the direction of an intruder, a posture similar to the vigilance posture as shown in Fig. 6.8. Others are meant for appeasement or submissiveness, as already described in Chapter 9. A distinct appeasement/submissive gesture by magpies, also used by currawongs, is the flutter of the wings involving not the entire wing, but mostly the primary feathers (called the manus flutter). Submissive behaviour in magpies is expressed by falling down, rolling over on the back and exposing the belly. One can see this behaviour within the play behaviour of juvenile magpies but also of juveniles when an adult pursues them. Usually, the juvenile will have done something wrong and the adult will pursue to punish but before being able to do so, the juvenile flings itself at once onto its back, at times emitting woeful sounds. The adult may still deliver a peck but usually the strategy works to appease the adult. Such use of body language is

Fig. 11.3. This juvenile (six months old) is play-fighting, displaying typical agonistic behaviour shown normally to intruding magpies: arched back, splayed wings, head arched forward and emitting sharp staccato 'call to arms' calls. There was no opponent and minutes later the same bird, sitting quietly on my arm, tried very subdued, charming begging calls.

rarely seen among birds, except parrots and corvids, but is much more typical of dogs and primates. It suggests that magpies have developed a particularly complex system of communication.

For agonistic interactions magpies may open their wings, as was shown in the magpie's play-fighting series or in squabbles over food already mentioned. In rare situations, play-fight posturing can take on very realistic forms (Fig. 11.3).

Feathers and facial expressions

In magpies, feathers are used for subtle signals during very close interactions. Signals, issued by feather posture alone, have rarely been systematically studied yet they may be quite important within visual range.[9] Feather movement and positioning is under the control of underlying musculature. Such musculature is essential in order to regulate moisture and temperature on the integument. Cold weather requires fluffing of the feathers to ensure better insulation. Feathers can be held erect to allow the sun to shine directly onto the skin. Magpies take sunbaths regularly and spread out their feathers to do so.

The same mechanisms of feather control can also be used in communication and express a surprising range of emotions. Magpies can fluff their feathers in a

Fig. 11.4. Affiliative cues. A hand-raised juvenile looking at the human 'parent'. Note the direct binocular gaze. Together with slightly extended feathers below the beak this is an expression of great tenderness and is usually accompanied by nearly inaudible purrs and tilting of the head. Sometimes, this posture is followed by allopreening (preening of the 'parent').

certain way when they are ill but they may also extend their feathers for other expressions.

Facial expression may also play a part in communication. The idea of a bird having a 'face' may seem odd – a face is something usually reserved for humans, or at least mammals, and is usually denied birds, with some exceptions.[10] Facial expression is achieved by independent positioning of feathers below or above the beak, on the flanks of the head, on top of the head (the crown), at the nape of the neck and, in some species, the feathers above the eyes can also move independently from other feathers. Magpies do not have a crest (as do cockatoos) but they can ruffle their flank feathers to express rising anger and possible attack. Sleeking of feathers is usually associated with fear but this commonly involves the whole body rather than just the head. Nestlings and even fledglings sometimes extend the feathers under the beak in a seemingly 'cuddly' and babyish manner (Fig. 11.4). These facial expressions are powerful signals emitted with a minimum of energy expenditure and in close encounters they may replace vocal signals altogether.

Rising anger can also be expressed very parsimoniously in a facial microsignal. The feathers above and below the eye can be raised to indicate displeasure (Fig. 11.5). They lie normally entirely flat against the skin.

Fig. 11.5. Facial feather expression. When irritated, magpies sleek down all their feathers but directly above and below the eye; feathers are raised, slightly in the first image and fully extended in the second image. It looks as if the bird has eyebrows, and note that the feathers under the beak are sleeked down. Once the cause of the irritation was removed, the feathers returned to their original position and the bird took no further action and not once vocalised. Non-verbal signals, i.e. signals that do not depend on voice, can thus be very important and may avoid conflict.[11]

Sound signals

Magpies have two types of audible signals that are not derived from the syrinx and these are used within the large repertoire of warning signals. One is beak-clapping. Beak-clapping in Australian magpies, as well as in storks, some owls, and in the various frogmouths are used as strong and aggressive warning signals. Interestingly, this signal is usually reserved for interactions with other species. Magpies may also use wing flapping and wing beating, (such as in wood pigeons and crested pigeons) as warning signals. When they are about to attack, they may produce an audible whistle in their wing beat but it is not clear whether this signal can be heard from some distance away.

Most auditory signals, however, are vocal and (as most Australians are aware) the magpie is one of the most vocal species. Vocalisations occur all year round, are issued by any age group (post-fledging) and by both sexes. This is in stark distinction to most birds in the Northern Hemisphere where the woods may resound with the calls of songbirds in spring – especially when males sing their breeding songs – but they are nearly silent from late autumn and through the cold winters when birds either migrate or fall silent.

A magpie vocalisation may not contain a message but, to a fellow magpie, it will reveal the identity of the sender. Voice recognition plays a very important part in birds living in colonies or groups. Parents may find their offspring by voice identification only.[11]

Beyond the general syntax (structure) of a vocalisation, there may also be a message (the semantics). Both can be intertwined. In socially living birds, as are magpies, it is efficient to communicate about several things because any communication may help to keep the group protected and assured of its territory. Vocalisations can inform about food, can express anxiety or alarm, rivalry, attention, defence, flying away (follow me) and similar short instructions. None of these may be specific to the sex of a bird. Males and females alike will utter calls in relation to predators and a host of other situations.

Vocalisations can reveal the territorial and broader geographic origin of a specific individual.[12,13,14] There may also be individual markers in a call or duet.[15] As dispersal is part of a magpie's usual life cycle it is not surprising that there are many variations and even dialects derived from territorial affiliations and regional peculiarities.[14] Dialects are, of course, known in many other avian species.[16,17,18] Song can also identify a neighbour because of some song sharing.[19,20] Neighbouring birds will know the territorial borders and therefore a form of truce may exist between neighbours. Finally, although this is not yet proven, there may be variations that identify the individual magpie.

Carolling

Carolling is a very specific and highly recognisable category of call in magpies. Short territorial calls are not uncommon in vertebrates but, in songbirds, it is usually the song itself that has territorial defence functions, saying: we are here and this is our territory. The song reminds conspecific neighbours or warns strangers that the territory is taken.[21] Magpies, however, have these special carolling calls for territorial defence. Some signals of this kind have also been found to identify the individual singer[15,22] and may have evolved as an energy saving and non-confrontational way to keep the peace or at least maintain an uneasy truce between neighbours.[23] Such observations have led to the 'dear enemy' hypothesis,[24] the idea that neighbouring territorial animals become less aggressive toward one another once territorial boundaries are well established. Magpie neighbourhoods are often very stable and magpies have been observed to share about 25% of their song repertoire with neighbours,[19] showing that a group 'belongs' where it is. However, there has now been plenty of research showing that the emphasis may sometimes be more on 'enemies' than on 'dear',[25,26] depending on the season.[27]

When all magpies in a territory carol together (chorus), the purpose is a bonding exercise after a successful mission of expelling a predator, for instance, and thus similar in function to the chorus laughter of kookaburras.

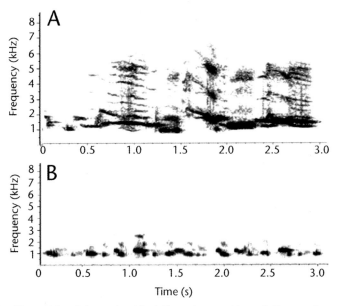

Fig. 11.6. A, Carolling and B, quiet song (warbling), the major sound types in the magpie repertoire. The carolling call shows strong energy even at 5 kHz; the grey elements indicate noise rather than clear sounds. The calls are strongest at 1.5–2 kHz and some individual calls even have harmonics (the parallel lines within the same column).

The magpie's territorial call has a very specific structure, is very loud and slurred and in stark contrast to the mellow, melodious sounds of their song (Fig. 11.6).

To appreciate the extraordinary quality of the caroling call, it is useful to compare the sound features with its normal quiet song, also called 'warble,' as represented in Fig. 11.6B. The difference in amplitude between the two types of vocalisations is very obvious. Carolling is loud, noisy and powerful. By contrast, the song has no overtones and little noise – just a gentle sway of sounds rambling along in a manner very pleasant to the human ear, like a well-tuned perpetual mobile. Australian expatriates in England told me that the sound of magpies singing always made them homesick because this, in a sense, was one of the most regular and uniquely Australian sound profiles that was part of their childhood.

We tested the strength and importance of carolling in the field as Carrick had done in the 1960s and 70s.[45] He showed that by just playing back carolling calls without the presence of the male issuing them, the territory could be held successfully. Testing this in the field is relatively easy, although enormously time-consuming. Kelly O'Shea undertook a project for her Honours thesis in our laboratory testing the importance of song and carolling in territorial defence. She had recorded carolling and song from neighbours and distant territories (strangers)

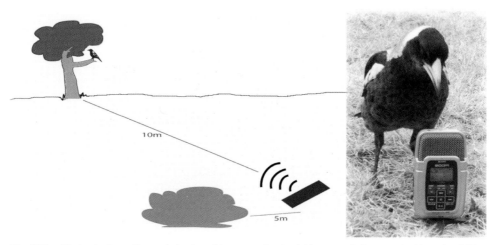

Fig. 11.7. Playback of carolling and short warble segments of neighbours and strangers inside and outside a magpie's territory. Vocal responses were swift. Only some juveniles were more interested in the small H2 digital recorder.

and then played back these recordings to designated groups inside and outside their territory (Fig. 11.7).

Duetting

Duetting occurs in many bird species and is known to play an important role in the vocal communication system of birds.[27] Duetting is a specific form of communication in which one bird initiates a call and another one answers. It usually refers to *sequential* calling rather than, as the term is used in human song, singing together – the latter is carolling. Duetting is useful when visual contact is lost or is at risk of being lost and may also synchronise territorial defence and seems to occur more frequently among prolonged monogamous pair bonds. Magpies duet regularly and so do magpie larks. Studies of zebra finches have found that pair-bonded birds engage in contact call duets far more frequently and in a non-random fashion than unpaired birds, hence showing that duetting, in such cases, also functions as means of bond confirmation and identification.[28] Duetting in magpies may have similar functions as in zebra finches, over and above the function of providing locational cues of callers. When adults come together to carol, we call it a duet when there is a pair of magpies and a chorus if there are more than two. However, the purpose of the carol is the same: a declaration of victory and/or a reaffirmation of the bonding of the pair or the group (Fig. 11.8).

A most unusual kind of duetting between two juvenile birds was observed and recorded that I believe to be unique. It was a form of duetting that did not just happen once but over several days and of a kind that I believe has never yet been

Fig. 11.8. Pair at the end of the day carolling overlooking their territory. Female started the carolling, male joined in and they performed several bouts of carolling together and overlapping and then flew off to roost for the night. There was no provocation or external stimulant observed. It is likely that such specific carolling bouts may be issued to express and reinforce bonding.

observed, or at least not reported. It was a duet between a juvenile pied butcherbird (*Cracticus nigrogularis*) and a juvenile magpie. The juvenile butcherbird seemed to fly in from a neighbouring territory (no adult butcherbirds were located in this magpie's territory). It was neither stopped nor harassed by the adult magpie group. It took up a position near a quietly singing juvenile magpie and then started vocalising, singing its soft song simultaneously and thus overlapping as if one wanted to out-compete/drown out the other. After a while, both birds stopped and then started again. This time, there seemed some deliberate adjustments to the sequences by both birds because they were now alternating song very closely and one seemed to wait for the other to intervene with the next element. Unfortunately, I had no recorder at hand when this chance discovery was made. However, I did get the field-recording equipment ready to use in case it happened again. Indeed, the next day the bird turned up again and as soon as I heard the butcherbird, the recording button was pressed and I managed to get close enough to capture the sounds as Fig. 11.9 shows.

Duetting between magpies, as was explained, always occurs in the form of carolling and has distinct social and territorial functions that communicate: 'we are here, this is our territory and we are a team'.

But this was duetting in song, and not in carolling, and it is doubtful that it had any broader communicative function. Butcherbirds and magpies, as was shown in Chapter 1, are close relatives. They even look somewhat alike, although the

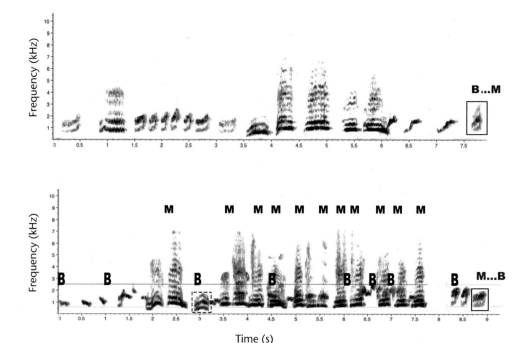

Fig. 11.9. A unique duet between a pied butcherbird and a magpie, both juveniles. B = butcherbird sound; M = magpie sound; B...M = butcherbird produces a sound similar to a magpie sound (convergence); M...B = magpie produces a sound compressed like the tonal element of the butcherbird song. The boxes indicate convergence, while the dashed box shows the butcherbird sound adapting to the magpie. Note the tall columns are each one sound and are emitted by the magpie. The darker lines indicate that the energy at that frequency level is the one that we hear (1st and 2nd formant). The parallel lines above the darker ones are harmonics, which are referred to as overtones, giving the sound a richness, very similar to the way a human voice produces overtones. The butcherbird, by contrast, has fewer or no overtones at all, making the sound appear more like a whistle.

butcherbird is smaller, and they share in being able to produce beautiful and varied song. I cannot explain this behaviour other than to say what I saw: they were deeply engrossed in their activity, had an erect and attentive posture and were very clearly closely attending to each other's vocal contributions.

For some time there has been a question over whether songbirds have an aesthetic sense of music appreciation and a good deal has been written about this in recent years.[29] Witnessing the duet, I was reminded of a report, published in *Nature* in 1903, on a duet by magpies based on the mimicry of a piece of music.[30]

Fig. 11.10. Mimicked and shared musical tune, shared between two magpies, each taking half of the tune in perfect timing.

The interesting part of this was that one bird sang the first part and the second one took over at the right time and concluded the segment (Fig. 11.10).

We know from the research on magpie memory and vocalisations that it is not at all difficult for magpies to learn and memorise specific sounds and strings of sounds but examples of duetting in song have remained exceedingly rare and remain without compelling explanations.

Alarm and distress calls

Alarm calls and distress calls are two call types both used within and between species. Nearly all birds and mammals have distress calls and it is more than feasible to assume that the signallers use the signal in an emotional, undirected manner. Fear produces high frequency sounds of relatively shrill tonal quality almost right across the animal kingdom. While others may decipher the context, distress calls are not necessarily calls for help by the sender and they can also be costly by promoting detection.[31] Still, from an evolutionary point of view such

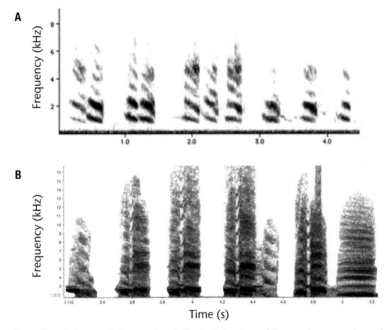

Fig. 11.11. Examples of alarm calls in magpies. A, Typical structure of the most common type of alarm call for relatively low level dangers. Particularly, the first three columns represent the most common generic alarm call (found in Melbourne, Canberra, Armidale and Coffs Harbour coast, thus involving at least two subspecies). B, A high arousal alarm call (8–10 kHz) usually uttered when first spotting an intruder and designed to attract the attention of conspecifics, expression of real arousal and anger. Note that the call is not only substantially louder than the generic alarm call but a good deal noisier (dark grey areas). Each column is a single sound, but also note the split in that sound which is audible as a strange click.

expressions of emotion are not without powerful communication content. Other birds may approach to assist and may drive away a potential predator or help save the rest of the group from further casualties.

Alarm calls may signal the presence of any type of predator and also the emotional state of the sender, so providing the receiver with information about imminent danger. We now have a good understanding of the alarm calls of many Australian birds, including the magpie.[32,33] Alarm calls that warn of flying predators are similar among very different species of birds and delivered with similar intensity and at about the same pitch of 6–8 kHz.[34] This pitch tends to avoid clear identification of direction of the call and, because of its similarity across species, the alarm call of one species may benefit a variety of other species.

The most frequently used alarm calls by magpies are the two represented in Fig. 11.11, one is generic, although the generic call outnumbers the high arousal call by 4:1.

I know what you are saying (referential alarm calls)

It is obviously not just a matter of how many calls a bird can make but the question is whether different calls can convey specific messages. I identified 27 different kinds of alarm calls in magpies and, of course, one wonders why a species would have so many if they all conveyed no more than a single type of call can. To simplify the list of alarm calls, calls were grouped together that seemed different only in small details but basically had the same parameters. Even when these were grouped together, there still remained six major groups of alarm calls.

One, already shown above (Fig. 11.11), was identified as a 'generic' alarm call, a stereotyped call for everyday events that might just say 'careful' or 'watch it' (Fig. 11.11A). Adults often address juveniles with a slightly shorter and lower amplitude version of this generic alarm call. Then there is an alarm call that carries with it aggression and signals attack (Fig. 11.11B), usually in conjunction with an intruding raven, bird of prey or a magpie group on an ambush to take over territory. One call was clearly identified as a high arousal alarm call (not shown) indicating that the caller was really experiencing something unusual. Still, this leaves three alarm call types unaccounted for.

Over the years, we conducted many field experiments with playback and with taxidermic models of predators to establish the range of alarm calling and eventually were able to identify two alarm calls that are referential in magpies.

Referential signalling refers to a form of signal that carries unchanging meaning and is seen as the beginning of lexicon, hence the first development towards language development. Referential signals tend to refer to predators, and even to different classes of predators, and sometimes to food. Evans and Evans[35] discovered that cockerels have a specific food call to attract females to come closer and, when they do, males sometimes use it not to advertise food but, deceptively, to mount the females. Cockerels were also shown to make a distinctly different warning call for a predator seen flying overhead than they do for one approaching on the ground.[36] There is a specific screeching call for an aerial predator and a clucking call for a ground predator. These calls are elicited only when there is another member of its own species nearby.[37] Without an audience, cockerels show the same amount of looking overhead, crouching and sleeking down of their feathers in response to seeing the overhead predator but they do this without alarm calling. In other words, cockerels issue alarm calls only when there is a reason for doing so and that, when they do call, they do not do so merely on impulse (expressing a state of emotion), but with the intention to communicate a very specific message.[38]

Our research found that magpies have a referential alarm call that we termed 'eagle alarm call' but it might as well apply to more than one type of bird of prey.[39] When we had finally isolated a call that seemed to be specifically designed to warn of a bird of prey, we then prepared tapes for playing back these very same calls to the magpies and then score their response. We played these calls back to magpies in over 30 territories, including to different subspecies and in very different locations, and mixed this in with other alarm calls and other vocalisations and then scored their response.

Results were very clear and unambiguous. The sound source for the playback was placed on the ground but the bird looked up instead. To explain this further – the playback contained randomly played single sounds, be these well-trialled alarm calls of the generic type, and calls that had small segments of the eagle alarm call in it as well, (called 'mixed') as well as the call we had thought to be a referential call (i.e. the eagle alarm call). Every call caused some behavioural reaction, mostly by raising the head, adopting a vigilant posture and looking around. At the very least, even if they had been feeding before, they stopped feeding.

These eagle alarm calls are probably among the most unusual ones one may ever hear uttered by a songbird. It is difficult to describe them because of their complexity. First, they are very loud and of a broad frequency range but with their main energy at high frequencies making these sounds carry above the frequencies in the environment. Second, they are continuous and third, they sound like small ringing bells on a sleigh ride through the snow, high, yet well balanced and tonal in quality. It is wall to wall sound and one, by its stark difference to any other ambient sounds in the environment, is indeed a most effective alarm call that alerts anyone within a wide area because its amplitude and frequency takes this call above other ambient noise. Fig. 11.12 is divided into three parts, the first showing an image of a magpie having detected a wedge-tailed eagle in the sky. The other two are sonograms. Fig. 11.12B identifies the individual elements and Fig. 11.12C presents a recording in natural time – showing a 'curtain of sound'. Note that the structure of

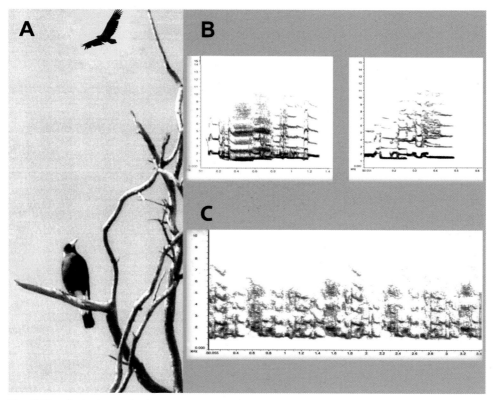

Fig. 11.12. Eagle alarm calls. A, magpie viewing the wedge-tailed eagle overhead – note when the eagle is so clearly visible in the sky, alarm calls tend to be brief presumably because other magpies will see it also. However, if it is more difficult to see, has landed or is immediately within the territory, the full alarm calls will be issued. B, Typical individual elements of the eagle alarm call, with its complex structure and high energy even at 5 kHz. C, 3.5 sec excerpt of the actual vocal performance of a single magpie's calling. Despite the high decibels, the sounds have relatively little noise and the wall to wall sound is like vigorous ringing bells.

the call is visually quite different to any other sounds magpies make – a kind of pagoda structure.

These were expected reactions knowing that magpies are very attentive to any different sounds and are likely to listen carefully and to investigate.

The real test was whether a replay of the alarm calls that we believed to be referential alarm calls produced a different response. As Fig. 11.13 shows, there was a strong and consistent reaction that remained the same in each of the 30 territories over a distance of more than 300 km tested.

Looking up while the sound-source was on the ground can only mean that the message was not to look for the owner of the sound but for the actual content of the message, such as 'look up – there may be an eagle in the sky'. Importantly, it seems that this signal is understood everywhere. We took the same recordings outback and found that magpies spontaneously looked up towards the sky.

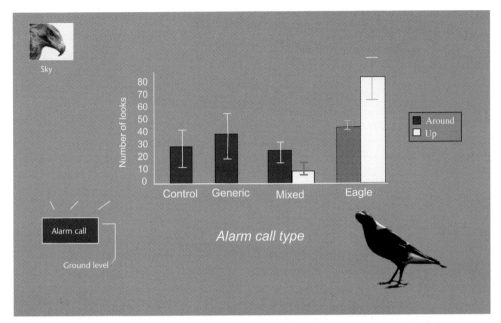

Fig. 11.13. Referential alarm call playback. Playback was conducted over a two minute period. The number and type of responses were recorded for four different calls (including one control) and then tabulated. For the innocuous control calls, little interest was shown and brief visual checks were conducted. The generic alarm call led to more sustained interest but exclusively of looking around. However, when the 'eagle' alarm call was played highly vigilant behaviour followed. The chief difference to other, and even mixed, calls was the strenuous effort in scanning overhead and the sky (last, white column). The grey column indicates movement of head starting with looking around but then also shifting the attention towards the sky. Note that the sound source is on the ground while the magpie scans the sky.

Unusually, we also had the opportunity to test the same calls on black-backed magpies on the Island of Taveuni, Fiji, north-east of the main island.[40] Palm oil plantations had become infested with palm moths and the mistaken idea was that magpies might feed on these insects and be a biological pest control. But, as was the case in the disastrous introduction of the cane toad to Australia, so it was in Taveuni. The introduction was entirely ineffective. The palm moths and the magpies never met, because magpies are strictly ground feeders while the palm moths live high up in the palm fronds. The first pairs of black-backed magpies were introduced around the turn of the 20th century and some later introductions of white-backed and black-backed pairs occurred mid-20th century. The magpies are struggling in Taveuni because of humidity and wet, as well as lack of trees (most native trees had to make way for palm oil plantations), but also because of an overwhelming presence of goshawks and Pacific (swamp) harriers (Fig. 11.14, see also Plate 16).

Again, we were able to show that magpies responded strongly and unambiguously to the eagle alarm call and looked at the sky.[40]

Fig. 11.14. Magpies and birds of prey in Taveuni, Fiji. For magpies, living in palm trees on Taveuni provided some protection from the weather but not from birds of prey. They were not only plentiful on the island but they are also ambush hunters and roosted in exactly the same areas as magpies. Magpies had to be extremely vigilant at all times to stay alive in the presence of very agile hunters. In the left image they were directly looking at a raptor in the next palm tree (middle image). The right image is a local swamp harrier.

More unusually, magpies have yet another way of communicating about a dangerous predator, such as a wedge-tailed eagle, by engaging in a symbolic behaviour that carries both vocal and visual information. Let us assume that a wedge-tailed eagle is no longer in the sky but has gone to ground, perhaps due to a previous hunt but hidden by some shrubs. I designed several experiments around this question using a taxidermic model of a wedge-tailed eagle and magpies showed a behaviour that had never been described in a bird and one that was said to be a cognitively complex behaviour described in children and great apes. This is pointing behaviour – a gesture that symbolically invites others to look, not at the person pointing but in the direction of the object or event to which the sender of the message wants to draw attention.

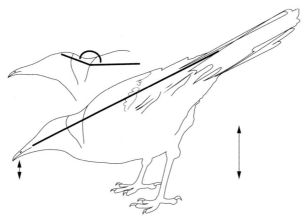

Fig. 11.15. Pointing posture. The extraordinary posture of the magpies while vocalising, sitting on a branch above the level of an eagle, pointing downwards, a combination of activities very difficult to maintain. Note that any forward position normally still involves an angle of the spinal vertebrae in the neck and back as shown in the upper excerpt. Instead, in the beak pointing posture the line is straight. An arriving magpie looks at the posture and follows the direction of body posture and beak and finds the eagle.

Gestures have rightly been thought of as being improbable in animals not equipped with arms. Chimpanzees have been shown to have extensive gestures and primatologists widely believe, and have evidence, that gestures may well be the first step towards language.[41] In magpies, a gesture would have to be done by body posture and use of neck and head in order to convey a sense of direction. Magpies used the 'eagle alarm calls' incessantly but, in addition, contorted their bodies forward in such a way that incoming magpies joined the pointing magpie and also performed the same action (Fig. 11.15), pointing correctly (despite changed angle) until everyone of the group was there and everybody had seen the eagle on the ground.[42]

This difficult and postural exaggeration is in line with beak and eye direction. Eye-gaze following is not quite the right way to describe it but may be in the same class of cognitive behaviours. Eye-gaze following plays a central role in social cognition and in the awareness of someone else's state of mind[43] – it is the ability to monitor and match another's head and eye orientation by following gaze direction into distant space. The beak pointing seems superfluous but the binocular vision of magpies, as was shown in Chapter 4, is only short range, so 'just looking' would not do but indicating direction, as precisely as possible, is. The importance of this is that the gesture and the calling have a strong facilitative role in creating effective joint attention[44] *and* action.

Overall, then, the communication system of magpies is extremely diverse and complex. Its qualities, including the ability to use referential signals, appear rare and thus very special. Australian magpies, moreover, are among the most superb songbirds in Australia and perhaps even the world. They can sing over four octaves, modulate complex sounds, have rich overtones or flute-like qualities, strike two tones at once and use an extremely wide range of sound structures. They are excellent communicators, as befits their territorial needs, but they are also first-rate musicians. In lyrebirds, mimicry has a concrete function. In magpies, we are not sure.

The palette of magpie vocal and non-vocal signals is certainly rich and varied. While ecological pressures might well explain their evolution, we can rightly say that the magpie makes a very good model species for the study of the complex relationship between social cognition, longevity and cooperative group living, as magpies are in specific contexts.

Endnotes

1. Anderson *et al.* 2017; Chen *et al.* 2016
2. Mirville *et al.* 2016
3. Ridley *et al.* 2005
4. Vallentin *et al.* 2016
5. Fischer *et al.* 2014
6. Panchanathan and Frankenhuis 2016
7. Kaplan 2014
8. Busnell 1977; Slabbekoorn 2017

9. Morris 1956; Kumar 2003
10. Bateson *et al.* 1980
11. Hinde 1972
12. Emlen 1972
13. Treisman 1978
14. Brown and Farabaugh 1990
15. Brown and Farabaugh 1991
16. Jenkins 1978
17. Baker 1983
18. Williams and Slater 1990
19. Farabaugh *et al.* 1988
20. O'Loghlen and Beecher 1999
21. Farabaugh *et al.* 1992
22. Falls and Brooks 1975
23. Fisher 1954
24. Temeles 1994
25. Christensen and Radford 2018
26. Courvoisier *et al.* 2014
27. Hyman 2005
28. Blaich *et al.* 1996
29. Taylor 2011, 2017
30. Waite 1903; Kaplan 2009
31. Bergstrom and Lachman 2001
32. Jurisevic and Sanderson 1994, 1998b
33. Wood *et al.* 2000
34. Marler and Tamura 1964
35. Evans and Evans 1999
36. Evans 1997
37. Marler and Evans 1996
38. Rogers and Kaplan 2000
39. Kaplan *et al.* 2009
40. Kaplan and Rogers 2013
41. Liebal and Call 2012
42. Kaplan 2011, 2015
43. Meltzoff and Brooks 2007
44. Itakura 2004
45. Carrick 1963

12
Magpies and humans

Magpies can be considered a delight and at other times a 'wildlife problem'.[1] There are points of friction that may stem from this bird–human interaction and there are many untold pleasures in having them around us.[2] Magpies, like most native birds, are generally healthy and disease-free. If magpies come to any misadventure, it is often inflicted or induced by humans. Conversely, a statistically miniscule amount of damage to humans has also happened over the years – minute even when compared with accidents such as falling off a ladder at home. Injuries sustained by cyclists, for example, are mostly the result of an accident caused by a fright from a swooping magpie rather than a magpie 'attack'.[3,4]

Some people have concluded that such mobbing or swooping behaviour shows that magpies are aggressive and even dangerous. A counter-argument would first point out that such swooping behaviour, as in countless other native bird species, occurs only during the breeding season and is usually confined to 4–6 weeks per year. For the rest of the year, magpies do other animals in their territory a favour by driving out predators and there are no clashes with humans at all. And in most cases there are also no clashes with humans during the breeding season. Magpies often set up territories in close proximity to people's properties. They get to know the owners, and even their pets, and never swoop them even at the height of the breeding season. This occurs in more than 80% of all people–magpie interactions, proving beyond doubt that magpies are not an aggressive species. If this were so, the birds' behaviour would be uniformly hostile all year round. Instead, they do not swoop people they know because, in their assessment, the ones that are nice to them, or even just indifferent, are assessed as posing no threat.

Direct attacks on the head, face and particularly on eyes have been noted in rogue magpies – a very small percentage of male birds that seem to have learnt to attack humans, regardless of their breeding status – and in a few isolated instances, eye injuries have required medical attention. Between 1986–1994 (an eight-year period) a survey containing 700 000 records of animals causing human injury and accidents in Australia identified 59 cases (less than 0.001% of all accident records) in which magpies were a factor in causing injury.[3] Reports of magpie attacks recorded by city councils are much higher than the recorded medical and hospital reports and many 'attacks' reported to city councils are not attacks because they involve no physical contact and no injury. In some specific urban places, reports of magpies swooping can reach hundreds in a season and hundreds are not necessarily hundreds of birds but a few birds going for every passer-by. In this way, a single magpie male can get quite a tally over a few weeks.

Conversely, mapies are often injured by humans. The largest collection of magpie injuries results from road accidents, which are usually fatal to the magpie if the impact is direct. Occasionally, birds are injured because they are caught in barbed-wire fences or they fly against a window. Poisoning as a result of pest control is another category of human-caused injuries to birds. The poisons used are commonly not intended for native birds but for foxes and wild dogs or they belong to the group of herbicides and insecticides that may sometimes be sprayed excessively in gardens.

Road accidents, injuries and disease

The victims of injuries from contact with cars do not usually belong in the category of the weak, sick, or old, as some opponents to rehabilitation are quick to argue. Rather, the victims tend to be the healthy ones. Those birds that are clipped or thrown forcefully by the air turbulence created by a passing car sustain broken limbs and bones, concussions, lacerations – indeed the whole range of injuries that humans may also sustain when hit by a car.

A preliminary investigation by members of the Wildlife Information and Rescue Services (WIRES) has found that the total number of road kills per state per day may number in the thousands. From the road kills on a limited number of rural roads they had counted, they estimated by extrapolation to the whole state that New South Wales alone would claim ~7000 victims of native animals daily. Magpies and kangaroos were at the top of the list of road kills of birds and of marsupials respectively.

Health statistics on wildlife in general are very patchy. We know even less of the health status of native avian species, let alone of magpies. Nevertheless, wildlife

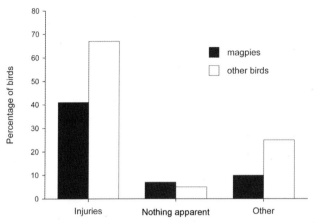

Fig. 12.1. A summary of the health profile of all birds taken into care by WIRES members in the New England Tableland in the years 1996, 1999, 2001 and 2002 needing medical or other attention. In all categories, except abandonment, magpies seem to require less assistance than other bird species.

rescue organisations such as WIRES have kept records and these can be used as a valuable resource to provide at least some evidence of the specific nature of problems for individual birds. Data from WIRES at the New England Tableland, counting the number of admissions of birds over four complete years (1996, 2000, 2001 and 2002), show that of the 1542 birds that were rescued, 351 were magpies (23%). Hence, almost a quarter of all birds requiring some kind of assistance were magpies. Even anecdotal evidence points to the conclusion that many magpies are killed on country roads following direct impact hits with cars (Fig. 12.1).

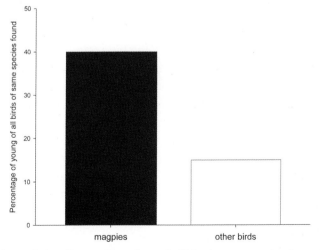

Fig. 12.2. Abandonment of nestlings and juveniles ranks highest among magpies. Some abandonment is the result of parent eviction (in case of fatal disease), others seem to try and fly too early and get lost and others are blown off the nest.

Fig. 12.3. This magpie lost its leg as a result of it being caught in binder twine, fishing line, or nylon string that had been incorporated into the nest structure. This bird actually survived the ordeal.

The figures indicate, however, that road accidents form the single largest category and the reason for hospitalisation (70%). Usually, the site of rescue is at or near a road, suggesting accidental trauma.

The category 'Other' in Fig. 12.1 refers largely to attacks by cats and dogs or to natural events. In songbirds and other small native birds cat attacks feature highest among rescue needs after car accidents, but this figure could be understated because these refer only to reported cases. It is well known that human cat-keeping habits and lenient cat-management policies contribute substantially to the decline of native birds. Magpies tend not to be affected by cat and dog attacks but they may suffer from injuries caused by wire or binder twine (see Fig. 12.3).

The least significant cause of health problems concerns disease or parasites. Indeed, the data suggest that this is a minute percentage. The category 'nothing apparent' refers to birds that could be taken with ease because they were grounded and unable to escape but they had no visible injury of any kind. In this category, we find all diseases, internal and external parasites, poisoning, nutritional deficiencies and other internal events. As can be seen, both for magpies as for any other bird species, this affects only a very small proportion of birds in general.

Nestlings and natural attrition in magpie populations

Health issues for nestling and juvenile populations differ from those for adults. For young birds, the primary issue does not relate so much to human impact but to the natural attrition rate of birds and introduced predators. This is particularly so for magpies. In the WIRES data, magpie nestlings and juveniles account for a staggering total of 40% of all magpie rescues (see Fig. 12.2). This is a very high figure for any bird species. On average, the young constitute ~15% of all birds of a particular species needing rescue. The reason for most rescue cases involving young magpies was separation. Some of the separated nestlings may have fallen from the nest at a time of severe storms; some were juveniles (often branchlings) that may have been carried away by strong winds and lost their way; some, as described above, were afflicted by injury or other ills.

These examples illustrate why the successful rearing of young magpies to flight stage is relatively low in magpies. For a variety of reasons, young magpies face several difficulties in growing up, making their survival to adulthood precarious.

Preparation for breeding

This book has extensively referred to the key notion of territoriality in magpies and how central it is to their survival. It may not have been noticed that throughout the entire year the magpie is a very peaceful bird, even friendly and approachable and never causes trouble for anyone – unless the 'trouble' is that the song is disturbing someone's sleep; or magpies that have befriended householders have learned to knock on the window a tad too early for a morsel; or the juveniles beg loudly early in the morning under someone's bedroom window. There is nothing in magpies, apart from their charm and readiness to make friends with people (more of this below), that could possibly even provide a hint that they have and could be 'trouble'. Incomprehensibly, *National Geographic* recently claimed that magpies are 'dangerous' birds by including them in its documentary *72 Dangerous Animals: Australia*. Dangerous animals are typically large, carnivorous or venomous and exert their biological advantages all year round. One learns very quickly when in tiger country in India or out in the bush in Africa how defenceless and exposed one can be and how high the risks are by standing at a river's edge full of crocodiles or hippos. Danger is what the animal itself represents and the size or skill at which it can assert its position and overpower, maim, kill or even eat humans. Most of them are predators. However, even large ungulates can become dangerous and smaller mammals such as dogs and kangaroos, emus and southern cassowaries can use their legs or teeth effectively to ward off intrusion, risk or danger if the need arises. Apart from an oversupply in venomous creatures, Australia is truly blessed in that we can freely hike in the bush without a disgruntled bear, moose, leopard, lion or tiger going into attack mode. My memories of southern Africa include a rather terrifying moment when I found myself suddenly standing in front of a wild rhino that had been entirely invisible in the scrub ahead. There was also the moment in the Bornean jungle when I came face to face with a fully grown wild orang-utan male. He could have killed me with one swipe of his powerful arms. These are humbling experiences. In most countries people have learned to live with such wild neighbours and accept their dangerous nature which, of course, also has a function and purpose in their environment. Take the top predators away, and entire ecosystems can decline.

Part of the reason why the magpie, a small and highly unlikely candidate for the fraternity of 'dangerous animals', is considered so is to some extent due to the way we use language. The word 'aggression' has slipped even into research articles when it is patently incorrect. Aggression is a value-loaded term that implies disapproval and alters and colours one's entire relationship to a person or an animal because of its moralistic overtones. We use such derogatory terms often and it gives us licence to judge.

In science and in relation to animals, especially when discussing birds, the term 'aggression' is applied incorrectly in almost all cases. There are very few – even exceptional – circumstances when the term 'aggressive' would be justified. The magpie is sometimes regarded as an 'aggressive' species. However, this is not correct – 'aggression' is a term borrowed from sloppy general use within human society. An example of 'aggression' is the so-called 'king hit', a one-punch attack on a person (who may even be a stranger to the one doing the punching) that results in seriously injuring or even killing that person. Aggression is a set of behaviours that serve no specific survival purpose and are non-functional or dysfunctional expressions of uncontrolled emotion. Aggressive acts are proactive strikes without there being an obvious cause and the terrible part is, as in king hits, it is bad for the recipient but usually also does damage to the person delivering it. By contrast, birds guarding their nests is a deeply functional behaviour shared by most species. What they do, they do to survive or protect their group or offspring.

In animal behaviour we speak of *agonistic* behaviour patterns, rather than aggression, when these have evolved to fulfil specific functions. Agonistic behaviour is adaptive for the entire species. It represents a well-evolved strategy to enhance the chances of survival either of an individual, a group or of offspring. It ranges from mild warnings to attack, usually in defence of offspring, a mate, a food source or a territory or all of them, as the need arises.

Nest defence is not 'aggression' but exactly what it says – a defence. It has nothing to do with 'anger' or 'hatred' and it is not an undirected, mean or deliberately nasty behaviour. Indeed, it is one of the most preserved behaviours in the evolution of species and found in all classes and orders. Almost all Australian nesting birds, at one time or another, engage in mobbing behaviour, telling animals and people that they don't want them to come any closer. The feisty little eastern yellow robin (*Eopsaltria australis*) does as good a job at that as any of the fantails or the ground-nesting lapwings and they all do so for an anxious 3–5 weeks in the entire year and then they calm down and get too busy trying to feed their young to bother about nest defence.

Magpies can be persuaded not to mob a passer-by

Magpies are most unusual in that one can persuade them to stop mobbing. It may not apply in all situations and there are specific sets of circumstances that make this less likely but, more often than not, the mobbing behaviour in the male can be switched off in an instant. I know of no other species in the bird world (no doubt they exist) where one can stop this most natural behaviour simply by being nice and non-threatening.

The reasons for this are relatively simple but also unexpected and very remarkable indeed. We tested this in various forms and discovered that magpies recognise faces and individuals and can and will remember individuals for years to come and whether they were nice, bad or dangerous. The point is that it all has to do with risk assessment and with their cognitive capacity to remember potential risks people may pose. Magpies form memories of 'enemies' as well as of friends, an ability that has also been documented in the American crow.[5] Many people befriend magpies in their backyards and in these situations attacks are usually unheard of.

When magpies decide to breed they survey their territory and take careful note of who is inside that territory. As was shown throughout this book, the daily attention to detail, the watchfulness to keep the territory safe, the routines and the careful monitoring of the territory, even the mimicry of sound elements within the territory show a preparedness for raising a family that even many humans may not match.

These days, magpie territories may often include humans. If the family is indifferent, civilised or nice and doesn't incite the dog to chase after birds, chances are that the human residents will only know of 'their' magpies having bred successfully when the familiar begging calls of youngsters appear. None of these human residents will ever be mobbed, not once. If the human residents have befriended the magpies or been there for many years, it is likely that the adults will introduce their youngster at the back porch and more: parents will allow the youngster to be talked to and there will be no cause of friction. Many people have formed long-lasting friendships with 'their' resident magpies because they will stay, if they can, for all their lives and that can be several decades long.

Indeed, magpies extend the same courtesy to walkers in public parks if they are regulars, just people walking, jogging or exercising their dogs. Even those, relatively transient, people the magpies will recognise and consider non-dangerous and thus dispense with any warning swoops.

The purpose of swooping on people

At mobbing events at nesting time, typically the magpie male (the female is brooding) will swoop to warn that a person or persons unknown have come too close to the nest. The emphasis is on warning. If a recipient of this warning – the human intruder – responds favourably (i.e. takes note of the warning), the magpie male will cease pursuit and, if the same person comes past again, will get free passage. The system works successfully even for cyclists if they ride slowly and are known to the magpies. Interestingly, a video clip once placed on YouTube showed locals in a cul-de-sac who had befriended the resident magpie family and nobody was mobbed. Trying out different forms of helmet decorations, one resident rode up and down the street and was mobbed every time, no matter whether he wore eyes on the back, metal markers that bobbed up and down as a deterrent, or used a different coloured helmet. When he finished his rides, he took his helmet off and the magpie stopped himself nearly in mid-flight. Again, the explanation is quite simple. Without the helmet, the magpie recognised him immediately and stopped the pursuit.

There are sometimes hotspots where mobbing behaviour is persistent. One can almost (and sometimes without fail) trace back such increases in conflict with magpies to human behaviour that was hostile and quite physical. There are people who think that using an umbrella and fencing a magpie will teach the magpie to keep its distance. While this may be intuitively justified, as a strategy it is achieving the opposite: magpies assess risk and fencing with an umbrella is obviously perceived as a hostile act and that human is then reclassified as 'dangerous'. That is, as in international diplomacy, de-escalation and escalation of a conflict may hinge on a few crucial responses and behaviours of recognition of the purpose of the warning swoops.

Moreover, swoops are not designed to attack, i.e. not to make contact. In flight, that could be very dangerous for the magpie because impact could break his neck.

Children are a separate category for various and complex reasons. Children chasing magpies would be classified high risk, children throwing stones at magpies might set off a chain reaction, as it happened in one case that I watched. The boys were in school uniform and threw stones at magpies, unfortunately not just as a sport but to injure and even kill the birds. A school bus took them away and the agony was over for the magpies. However, from then on, that school uniform became a call to arms for the magpies and many totally innocent children in the same school uniform became a target for very concerted magpie attacks. This bad outcome can also be turned around but not without some intervention. Without exceptionally bad behaviour by children or adults, most situations can be renegotiated with magpies and often may take just a little mincemeat to persuade the male that people want to cause them no harm (Fig. 12.4), bearing in mind that these conflicts are not all year round but literally affect only a 4–5 week window. Moreover, very few magpies make

Fig. 12.4. An adult male in breeding condition being persuaded into a truce by an occasional small supply of fine mincemeat. While wary and keeping a very close watch on the human provisioner of food, no overt hostile acts were ever observed.

it to breed successfully and increasingly it will depend on human attitudes whether the magpie populations continue to decline or remain stable.

Sometimes these situations fought out between local communities and councils can become quite bizarre. I was involved in a case where a council had received a complaint about a magpie and decided to act at once by removing the nest. The nest, containing three nestlings, was taken to a veterinarian who was charged with the task of killing these birds – a protected native species under the law. The veterinarian accepted the birds but refused to kill the healthy nestlings and asked whether I could raise them. I agreed and the youngsters, including the intact nest, were delivered to my home.

The birds were successfully raised and eventually released. There was just one problem. It was at once apparent that these nestling birds were not magpies. The inside of a magpie nestling's beak is very red. However, the inside of the beaks of these nestlings was a clear and bright yellow. Only butcherbird nestlings (while their head and beak shape and colouration may look like those of the magpie) have yellow gape. So the council officers charged with removing the magpies had instead collected a pied butcherbird's nest. Is there a moral to the story? The reason why the magpies had started swooping and were considered a 'nuisance' was the proximity of the butcherbird nest (just in the next tree). Butcherbirds, especially at breeding time, are not a good neighbour to have when nesting because they steal eggs and take live nestlings to feed their own brood. The butcherbirds represented a real risk to the magpies and intolerance to their close proximity was more than justified. Once they had gone, the magpies calmed down completely and quietly raised their young. Council felt good for their decisive action, the vet had done the right thing by not killing local wildlife willy-nilly, but the chain of reasoning was of course anything but biologically relevant or environmentally savvy. I suppose we still have a lot to learn to live peacefully with our own wildlife. On balance, as was stated in the introduction to this book, Australian magpies still rose to the status of Australia's most popular bird.

It is important to distinguish between 'swooping' and attacking. Generally, the word 'attack' is used but this often does not accurately explain the purpose of the magpie's approach. Magpie swooping is usually not meant to make contact or harm the person but to encourage the person to make a wider turn away from the nest. The sign that Brisbane City Council has erected is generally useful advice (Fig. 12.5).

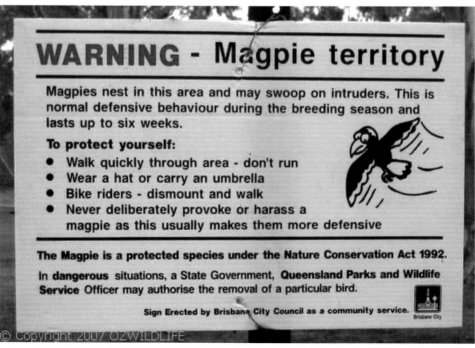

Fig. 12.5. Good road warning sign helps passers-by to deal with nesting magpies. It should have added perhaps for pedestrians and bicycle riders to move deliberately to the other side of the road. More councils should have signs like this that provide leadership of the right kind of behaviour and reassure the public.

Magpie defense usually follows several stages. In stage one, the magpie might beak-clap and swoop towards an object of intrusion but not near it. These are *warning* signals and, by themselves, do *not* constitute an attack. Then, if the intruder does not respond to these warnings, the magpie will resort to stage two and will swoop again, but this time closer to the target. Finally, and only if all earlier strategies did not work, the magpie will actually set out to attack.

It is clear from some individual birds that they may, in fact, leave out stages one and two, omit any warning signals and attack straightaway. Some magpies seem to have learnt over time that humans ignore their polite requests and warning stages completely, and that displaying these early signals is pointless. In busy cities, the magpies' judgement is correct, of course, so through our own ignorance and/or lack of response to their efforts of communication, we may have inadvertently conditioned a few of them to attack rather than to warn, to seek physical contact rather than swoop near us.

It is also true that there are now certain flash points for magpie swoopings, as magpie alert websites show,[6] which tend to be where traffic is high and there is a high turnover of pedestrians. In those cases, magpies are absolutely overwhelmed and stressed because all their careful risk assessment fails. The question is, why do they stay? The most important consideration for a magpie looking for a territory is to find a suitable nesting tree. The simple truth is that the number of suitable trees in urban areas is shrinking at an alarming rate, making any tree prime real estate, even if other circumstances are less than ideal. General responses are that we blame the bird and the bird has to be removed or killed in order to make a particular passage, a park or other public place safe again.

The introduction of compulsory helmet wearing for cyclists in NSW in 1994 and in other states may have been a good general safety measure. However, it may also have contributed to increasing the number of magpie attacks. When riders wear helmets, magpies are no longer able to tell who is riding the bike. Anonymity via helmets, as via balaclavas, raises immediate suspicion in magpies and humans alike because one cannot tell people apart individually.

An interesting variant of the hat/helmet-wearing problem was told to me during an ABC talkback show on native birds. The mother, Dee Hartin, said that her 9-year-old son regularly rides his scooter and interacts with a magpie but has never been harmed. Tylah puts his hat on and greets a waiting magpie, who accompanies him to the end of the street, waits for him and then flies next to Tylah on his way back home.[7] I asked the mother whether he ever took his hat off or put it on while the magpie was watching and she confirmed this. The magpie had seen him with and without the hat and instantly recognised him. It also helps that a hat does not hide the face as much as a helmet. In such a case, even the hat was no hindrance to a friendly relationship. The mother had also very sensibly told Tylah that the bird meant no harm (so he did not wave his arms around his head to try to shoo the bird away) and the bird had obviously drawn the same conclusion about the boy.

Birds, and most animals generally, do not have the ability to change their outer coat or plumage as we do with our clothes. With humans being so variable in appearance, it is not surprising that facial recognition has become so important to magpies.

Dealing with magpie mobbings

When we are considering how to deal with magpie attacks, there are some fundamental principles to do with magpie behaviour and survival that might be considered.

- Breeding magpies are the high achievers of magpie society. They have survived years of hardship, fought hard to get a territory and have been able to find a partner. The magpies that breed are healthy, mature, experienced and possibly the best stock we have.
- The main task of male magpies during the breeding season is to help feed the offspring and to defend the territory or, more specifically, the area around the nest site. The female also feeds but she has usually no time to take part in defending the nest site and, not surprisingly, research has found that 97% of all magpies swooping humans are male.[8]
- The quality of parents – their state of health and experience – will largely determine the survival of their offspring and the ability of their offspring to reproduce at a future time.
- The alertness of magpie males while defending nestlings is based on, and partly determined by, the 10-fold increase of gonadal testosterone, timed to aid in nest defence, not 'aggression'.[9]
- Finally, magpies often mate for life. Pair bonds tend to be intensely cooperative and mutually supportive and forceful separation may have poor outcomes for both birds.

In some parts of Australia, the current official government practice in cases of magpie attacks on humans is to kill or remove the offending magpie male. Killing or relocating magpie males during the breeding season means that the fittest, healthiest, most experienced birds are removed, that pair bonds are destroyed and offspring are left to an uncertain future. Removing magpies that have caused injury and releasing them elsewhere (translocation) usually leads to one of two disruptive outcomes: the birds may come back within a few days or weeks (their homing ability is as good as that of most flighted birds) but in the meantime its nestlings may have starved, the female may have been driven off and the nestlings abandoned, or its position may have been taken over by a less suitable subordinate male or the translocated bird, remaining separate like its mate and offspring, might be cast into an uncertain future.

Vacancies created by the removal of breeding males have sometimes been filled by so-called 'non-aggressive' males,[8] giving the impression that magpies adapt well to such family disruptions. However, this appearance is deceptive. The incoming 'non-aggressive' male is unlikely to engage in the alert and defensive behaviour necessary to protect a brood. Its 'non-aggressive' behaviour will probably be detrimental to the brood's survival.

Hence the removal of magpies during the breeding season remains controversial. In some bird species, it has also led to infanticide of the nestling.[10] More importantly, removal makes little sense as a form of biological management because the birds are tied to territory.

Friendships with magpies

As I mentioned in the introduction, there is overwhelming evidence that free-ranging magpies can form long-lasting friendships with people, which are enriching and harmonious for both the birds and humans. Friendships between humans and magpies also develop when injured magpies, which are usually found in the backyard, are rescued and cared for.

Fig. 12.6. Two juveniles resting in play, feet quietly locked and looking at each other. Photo: The Magpie Whisperer, www.magpieaholic.com.

Personal contacts with magpies reveal their behavioural repertoire and the many enchanting and complex qualities that they possess. We know from such encounters that magpies can be very gentle and tender and readily form friendships with humans – aspects that can be very touching. Magpies excel in playfulness, adaptability and inquisitiveness and they certainly do seem to have 'minds of their own'.[11,12,13] Magpies live in a well-ordered society defined by etiquettes, rules and hierarchies, where rewards and also punishments are quickly at hand should an offence occur. Magpies are capable of forming firm associations with human beings but they will do so only on being given repeated signs of goodwill.[11] It is a relationship that needs to be reciprocal and may require some work but it can be a long-lasting positive investment of living harmoniously with the magpie. These are not just my experiences, there is documented evidence of the rare salutary effects of some of these unusual friendships. [14,15,16]

Over the last 25 years, I have hand-raised dozens, treated scores of adults and spent time in the field with them observing their daily lives. Admittedly, to this day, there are still occasions when I am taken aback by the smart or charming or funny instances I've witnessed or of which I was made the subject. For one magpie, any quiet moment in the garden turned into a vigorous engagement with my shoe-laces, any hanging up of washing might result in a request for a fast turning of the Hill's Hoist so the magpie could spin around. I might have had the company of the only magpie in the world who has shouted at me 'I have got dinner for you'. Another experience with a magpie, one that had a broken wing and who needed exercise, led me to take this magpie on walks. Little did I realise that the exercise was all mine while the magpie managed to trick me. On the second walk around the property, waddling magpie in tow, he suddenly disappeared and I found him at the end of the circle. He was sitting there waiting for me, having taken a shortcut!

The large number of magpies from different groups and locations I have had the privilege to get to know has convinced me that my experiences with these birds make the birds, not my experiences, extraordinary.

Endnotes

[1] Jones 1996
[2] Jones and Thomas 1998
[3] Kreisfeld 1997
[4] Jones 2002
[5] Marzluff *et al.* 2010
[6] Phelan 2018
[7] Farrow-Smith *et al.* 2018
[8] Jones and Finn 1999
[9] Armstrong 1996
[10] Bekoff 1993
[11] Rogers 1997
[12] Kaplan 2015
[13] Meredith 2017
[14] Bloom and Greive 2016
[15] http://www.magpieaholic.com
[16] Kaplan 2017b

Epilogue: The success of magpies

There is no doubt that the Australian magpie is a very successful bird at many levels. They have learned to survive well as territorial owners, to communicate in complex ways and maintain a disciplined group of members that will work closely together as a team. Several feeding innovations and their extensive play behaviour also suggest that they are particularly smart. The magpie's impressive range of social activities, its willingness to interact with people, and its propensity to invent even leisure time activities, have made the magpie almost accessible company. Despite some misdemeanours, it has become a nationally significant and much-cherished bird. Its exceptional range of vocalisations and its familiar soft songs have made us listen.

The apparent success of magpies has inevitably led to suggestions that they have benefited from European settlement and Western agricultural practices that have given them more feeding grounds and better feeding options. It seems unlikely though that the magpie owes European settlers a debt of gratitude.

Survival of any native species in these changed conditions is not necessarily the result of deriving 'benefit' from human technological and farming expansion. What this rather suggests is that some species are more resilient than others. They have managed to replace former food sources with new, and often exotic, ones. The magpie has certainly shown it can adapt, as have most of the members of the artamid family.

Magpies, like many Australian songbirds, are generalists and can survive on vastly different types of food. They take whatever is available and manage to find enough food, regardless of season or abundance. It is likely that their wide repertoire of feeding options has contributed to their success and survival (see Chapter 5).

Australian native birds that are very specialised feeders tend to rank high on the list of endangered species, among them the glossy black-cockatoo, many nectar feeders and many birds of prey at the top of the food chain. The magpie's style of foraging may have also helped it adapt to the changing landscape.

However, we do not know whether the magpie is more abundant now compared to, say, 200 years ago. We simply have no data on this. It has been shown that there have been increases in range (and local increases in numbers), but this is an issue of range not necessarily of overall numbers.[1]

The introduction of the Australian magpies to other parts of the Pacific has generally not been overly successful. Magpies introduced to the islands of Fiji in the 20th century – about the same time as they were introduced to New Zealand – have now almost died out and those that remain in a few isolated spots are reportedly not doing well. In New Zealand, where they are still widespread, the mean life expectancy of magpies is said to be a mere six years,[2] a substantially shorter life compared to that of Australian mainland populations who live for 20 to 30 years. It would be of great interest to learn why Australian magpies have such vastly different life spans in these two countries. In Australia, magpies are widespread and found in most parts across the continent. They have shown a remarkable ability to adapt to different climate zones. They appear to be well organised, well fed and generally have few diseases. Altogether this indicates adaptability, success and, hopefully, a promising future for the species, if we let them. According to one Dreamtime story related in the first chapter of this book, magpies were holding up the sky, the least we can do is save a few trees for them and treat them well.

Endnotes

[1] Robinson 1956
[2] Heather and Robertson 1996, 'Magpies', p. 432

Plate 1. Beautiful magpie: understated, elegant, sensitive, intelligent and courageous.

Plate 2. Magpie in autumn.

Plate 3. An example of a colour variation in a juvenile magpie.

Plate 4. First-year fledgling with a mottled/scalloped breast.

Plate 5. Branchling looking very self-satisfied. This bird is barely over 4 weeks old but very adventurous. Note that the tail feathers are missing due to an attack by a raven when it was a nestling. Those tail feathers will not regrow for some time and initial flights attempts may be less successful but, in general, most birds can adapt to flying without tail feathers.

Plate 6. Singing magpie, barely opening the beak.

Plate 7. Solo play – moving in and out of the water and getting comfortable. Source: The Magpie Whisperer, www.magpieaholic.com

Plate 8. Three magpies engaged in social play. Note, one magpie is using the foot of another bird to give it a swing. Source: The Magpie Whisperer, www.magpieaholic.com

Plate 9. Juvenile playing with bell (object play). Source: The Magpie Whisperer, www.magpieaholic.com

Plate 10. Magpie listening intently. Note the position of the head and the twisting of the body. This is quite typical of a magpie trying to identify a sound and/or its source.

Plate 11. Magpies chasing a sea eagle. Our research on magpie defence strategies shows that magpies pursue large predatory birds such as sea eagles in pairs, with one magpie on either side of the predator's head to reduce the risk of attack. This photograph has been modified by the addition of the magpie on the right to represent this defence strategy. Original photograph by Akos Lumnitzer.

Plate 12. Curious magpie.

Plate 13. Three magpies browsing at sunset in the Mallee. Source: Kim Wormald, lirralirra.com

Plate 14. Magpie perched on a tree at Walpa Gorge, Northern Territory.

Plate 15. There is no competition for food between partners and couples and the first one can feed quietly while the second bird looks on with great interest and some extended feathers, suggesting rising inability to cope with seeing the other feed (we would call this jealousy).

Plate 16. Magpies eking out a living amongst palm oil plantations, in Taveuni, Fiji.

References

Ahlering MA, Faaborg J (2006) Avian habitat management meets conspecific attraction: if you build it, will they come? *The Auk* **123**(2), 301–312. doi:10.1642/0004-8038(2006)123[301:AHMMCA]2.0.CO;2

Alatalo RV, Lundberg A, Glynn C (1986) Female pied flycatchers choose territory quality and not male characteristics. *Nature* **323**, 152–153. doi:10.1038/323152a0

Alley TH (1979) Diet difference in adult and nestling Australian magpie. *Corella* **3**, 4.

Alonso Y (1998) Lateralization of visual guided behaviour during feeding in zebra finches (*Taeniopygia guttata*). *Behavioural Processes* **43**, 257–263. doi:10.1016/S0376-6357(98)00015-1

Amadon D (1951) Taxonomic notes on the Australian Butcher-birds (Family Cracticidae). *Novitates* **1504**, 1–33.

Anderson RC, Searcy WA, Peters S, Hughes M, DuBois AL, Nowicki S (2017) Song learning and cognitive ability are not consistently related in a songbird. *Animal Cognition* **20**(2), 309–320.

Armstrong DA (1996) Territorial behaviour of breeding white-cheeked and New Holland honeyeaters: conspicuous behaviour does not reflect aggressiveness. *Emu* **96**, 1–11. doi:10.1071/MU9960001

Armstrong EA (1963) *A Study of Bird Song*. Oxford University Press, Oxford.

Armstrong EA (1965) *Bird Display and Behaviour*. Dover Publications, New York.

Arning L, Ocklenburg S, Schultz S, Ness V, Gerding WM, *et al.* (2013) PCSK6 VNTR Polymorphism is associated with degree of handedness but not direction of handedness. *PLoS One* **8**(6), e67251. doi:10.1371/journal.pone.0067251

Arnold AP, Nottebohm F, Pfaff DW (1976) Hormone concentrating cells in vocal control and other areas of the brain of the zebra finch (*Poephila guttata*). *The Journal of Comparative Neurology* **165**, 487–511. doi:10.1002/cne.901650406

Ashton C (1986) Predation of young bee-eaters by a magpie. *Bird Observer* **656**, 96.

Auersperg A, Teschke I, Tebbich S (2017) Physical cognition and tool use in birds. In *Avian Cognition*. (Eds C ten Cate and SD Healy) pp. 163–183. Cambridge University Press, Cambridge, UK.

Auersperg AMI, van Horik JO, Bugnyar T, Kacelnik A, Emery NJ, von Bayern AMP (2015) Combinatory actions during object play in Psittaciformes (*Diopsittaca nobilis, Pionites melanocephala, Cacatua goffini*) and Corvids (*Corvus corax, C. monedula, C. moneduloides*). *Journal of Comparative Psychology* **129**, 62–71.

Austin JJ, Parkin DT (1996) Low frequency of extra-pair paternity in two colonies of the socially monogamous short-tailed shearwater *Puffinus tenuirostris*. *Molecular Ecology* **5**, 145–150.

Australian State of the Environment Committee (2001) *Australia State of the Environment 2001*. Independent Report to the Commonwealth Minister for the Environment and Heritage, CSIRO Publishing on behalf of the Department of the Environment and Heritage, Canberra.

Australian State of the Environment Committee (2016) *Australia State of the Environment 2016*. Independent Report to the Commonwealth Minister for the Environment and Energy. Australian Government, Canberra.

Baglione V, Marcos JM, Canestrari D, Ekman J (2002) Direct fitness benefits of group living in a complex cooperative society of carrion crows, *Corvus corone corone*. *Animal Behaviour* **64**(6), 887–893. doi:10.1006/anbe.2002.2007

Baker AM, Mather PB, Hughes JM (2000) Population genetic structure of Australian magpies: evidence for regional differences in juvenile dispersal behaviour. *Heredity* **85**, 167–176. doi:10.1046/j.1365-2540.2000.00733.x

Baker MC (1983) The behavioral response of Nuttall's white-crowned sparrows to male song of natal and alien dialects. *Behavioral Ecology and Sociobiology* **12**, 309–315. doi:10.1007/BF00302898

Baker MC (2001) Bird song research: the past 100 years. *Bird Behavior* **14**, 3–50.

Baker MC, Mather PB, Hughes JM (2001) Evidence for long-distance dispersal in a sedentary passerine, *Gymnorhina tibicen* (Artamidae). *Biological Journal of the Linnean Society* **72**, 333–343. doi:10.1111/j.1095-8312.2001.tb01319.x

Balda RP, Kamil AC (1989) A comparative study of cache recovery by three corvid species. *Animal Behaviour* **38**, 486–495. doi:10.1016/S0003-3472(89)80041-7

Baldwin M (1979) Nest of the Australian Magpie, Australian birds. *Journal of the New South Wales Field Ornithologists Club* **14**(2), 40–41.

Ball GF, Casto JM, Bernard DJ (1994) Sex differences in the volume of avian song control nuclei: comparative studies and the issue of brain nucleus delineation. *Psychoneuroendocrinology* **19**, 485–504. doi:10.1016/0306-4530(94)90035-3

Bangs O, Peters JL (1926) A collection of birds from southwestern New Guinea (Merauke coast and inland). *Bulletin of the Museum of Comparative Zoology* **67**, 419–434.

Barker RD, Vestjens WJM (1990) *The Food of Australian Birds. 2: Passerines.* CSIRO Division of Wildlife and Ecology, Lyneham, ACT.

Barker FK, Barrowclough GF, Groth JG (2002) A phylogenetic hypothesis for passerine birds: taxonomic and biogeographic implications of an analysis of nuclear DNA sequence data. *Proceedings of the Royal Society of London B: Biological Sciences* **269**(1488), 295–308.

Barker FK, Cibois A, Schikler PA, Feinstein J, Cracraft J (2004) Phylogeny and diversification of the largest avian radiation. *Proceedings of the National Academy of Sciences of the USA* **101**, 11040–11045.

Barr J (1986) Magpie attack on sparrow. *Canberra Bird Notes* **11**, 133.

Barrett G, Silcocks A, Barry S, Cunningham R, Poulter LR (2003) *The New Atlas of Australian Birds*. RAOU, Birds Australia, Melbourne.

Bateson PPG, Lotwick W, Scott DK (1980) Similarities between the faces of parents and offspring in Bewick's swans and the differences between mates. *Journal of Zoology* **191**, 61–74. doi:10.1111/j.1469-7998.1980.tb01449.x

Battley PF, Warnock N, Tibbitts TL, Gill RE, Piersma T, Hassell CJ, Douglas DC, Mulcahy DM, Gartrell BD, Schuckard R, Melville DS (2012) Contrasting extreme long-distance migration patterns in bar-tailed godwits *Limosa lapponica*. *Journal of Avian Biology* **43**(1), 21–32.

Baumel JJ (Ed.) (1993) *Handbook of Avian Anatomy*. Nuttall Ornithological Club, Cambridge, MA.

Bedggood GW (1976) Recent feeding observations. *Bird Observer* **539**, 77–78.

Bednarz JC (1988) Cooperative hunting Harris' hawks (*Parabuteo unicinctus*). *Science* **239**(4847), 1525–1527. doi:10.1126/science.239.4847.1525

Bekoff M (1993) Experimentally induced infanticide: the removal of birds and its ramification. *The Auk* **110**, 404–406.

Bekoff M (2001) Social play behaviour. Cooperation, fairness, trust, and the evolution of morality. *Journal of Consciousness Studies* **8**(2), 81–90.

Bell HL (1980) Composition and seasonality of mixed-species feeding flocks of insectivorous birds in the Australian Capital Territory. *Emu* **80**(4), 227–232.

Bell BA, Phan ML, Vicario DS (2015) Neural responses in songbird forebrain reflect learning rates, acquired salience, and stimulus novelty after auditory discrimination training. *Journal of Neurophysiology* **113**(5), 1480–1492. doi:10.1152/jn.00611.2014

Bénard F, Callot JP, Vially R, Schmitz J, Roest W, Patriat M, Loubrieu B (2010) The Kerguelen plateau: records from a long-living/composite microcontinent. *Marine and Petroleum Geology* **27**, 633–649.

Bender H (2018) Caching or coating? An interesting behaviour by the Little Raven *Corvus mellori* in an urban setting. *Victorian Naturalist* **135**(1), 18–19.

Bennett ATD, Cuthill IC, Partridge JC, Lunau, K (1997) Ultraviolet plumage colors predict mate preferences in starlings. *Proceedings of the National Academy of Sciences of the United States of America* **94**, 8618–8621.

Beran MJ (2015) The comparative science of 'self-control': what are we talking about? *Frontiers in Psychology* **6**, 1–4. doi:10.3389/fpsyg.2015.00051.

Bergstrom CT, Lachmann M (2001) Alarm calls as costly signals of antipredator vigilance: the watchful babbler game. *Animal Behaviour* **61**, 535–543. doi:10.1006/anbe.2000.1636

Bernard DJ, Casto JM, Ball GF (1993) Sexual dimorphism in the volume of song control nuclei in European starlings: assessment by Nissl stain and autoradiography for muscarinic cholonergic receptors. *The Journal of Comparative Neurology* **334**, 559–570. doi:10.1002/cne.903340405

Bertram B (1992) *The Ostrich Communal Nesting System*. Princeton University Press, Princeton.

Birkhead TR, Pellat J, Hunter FM (1988) Extra-pair copulation and sperm competition in the Zebra Finch. *Nature* **327**, 149–152.

Black A (1986) The taxonomic affinity of the New Guinean magpie *Gymnorhina tibicen papuana*. *Emu* **86**(2), 65–70. doi:10.1071/MU9860065

Black JM (1996) *Partnership in Birds*. Oxford University Press, Oxford.

Blackmore CJ, Heinsohn R (2007) Reproductive success and helper effects in the cooperatively breeding grey-crowned babbler. *Journal of Zoology* **273**, 326–332. doi:10.1111/j.1469-7998.2007.00332.x.

Blaich C, Steury KR, Pettengill P, Mahoney KT, Guha A (1996) Temporal patterns of contact call interactions in paired and unpaired domestic zebra finches (*Taeniopygia guttata*). *Bird Behavior* **11**(2), 59–69.

Blakers M, Davies SJJF, Reilly PN (1984) *An Atlas of Australian Birds*. RAOU, Melbourne.

Bloom C, Greive BT (2016) *Penguin Bloom: The Odd Little Bird That Saved a Family*. ABC Books, HarperCollins, Australia.

Boland CRJ, Cockburn A (2002) Short sketches from the long history of cooperative breeding in Australian birds. *Emu* **102**, 9–17. doi:10.1071/MU01039

Boland CRJ, Heinsohn R, Cockburn A (1997) Deception by helpers in cooperatively-breeding white-winged choughs and its experimental manipulation. *Behavioral Ecology and Sociobiology* **88**, 295–302.

Boles WE (1995) The world's oldest songbird. *Nature* **374**, 21–22. doi:10.1038/374021b0

Boles WE (1997) Fossil songbirds (Passeriformes) from the Early Eocene of Australia. *Emu* **97**, 43–50. doi:10.1071/MU97004

Boles WE (1999) A new songbird (Aves: Passeriformes: Oriolidae) from the Miocene of Riversleigh, northwestern Queensland, Australia. *Alcheringa* **23**, 51–56. doi:10.1080/03115519908619338

Boles WE (2005) Fossil honeyeaters (Meliphagidae) from the Late Tertiary of Riversleigh, north-western Queensland. *Emu* **105**, 21–26. doi:10.1071/MU03024

Boles WE, Godthelp H, Hand S, Archer M (1994) Palaeontological note. *Alcheringa* **18**, 70. doi:10.1080/03115518.1994.9638764

Bolhuis JJ, Everaert M (Eds) (2013) *Birdsong, Speech, and Language: Exploring the Evolution of Mind and Brain*. MIT Press, Cambridge, MA.

Brenowitz EA (1997) Comparative approaches to the avian song system. *Journal of Neurobiology* **33**, 517–531. doi:10.1002/(SICI)1097-4695(19971105)33:5<517::AID-NEU3>3.0.CO;2-7

Brenowitz EA, Margoliash D, Nordeen KW (1997) An introduction to birdsong and the avian song system. *Journal of Neurobiology* **33**, 495–500. doi:10.1002/(SICI)1097-4695(19971105)33:5<495::AID-NEU1>3.0.CO;2-#

Brockie RE, Sorenson L (1998) An Australian magpie's (*Gymnorhina tibicen*) response to fake snakes in New Zealand. *Notornis* **45**, 269–270.

Brown C, Margat M (2011) Cerebral lateralization determines hand preferences in Australian parrots. *Biology Letters* **7**, 496–498. doi:10.1098/rsbl.2010.1121

Brown ED, Farabaugh SM (1990) Macrogeographic variation in alarm calls of the Australian magpie *Gymnorhina tibicen*. *Bird Behavior* **9**, 64–68. doi:10.3727/015613890791749055

Brown ED, Farabaugh SM (1991) Song sharing in a group-living songbird, the Australian magpie, *Gymnorhina tibicen*, III: Sex specificity and individual specificity of vocal parts in communal chorus and duet songs. *Behaviour* **118**(3), 244–274. doi:10.1163/156853991X00319

Brown ED, Veltman CJ (1987) Ethnogram of the Australian magpie (*Gymnorhina tibicen*) in comparison to other Cracticidae and Corvus species. *Ethology* **76**, 309–333. doi:10.1111/j.1439-0310.1987.tb00692.x

Brown ED, Farabaugh SM, Veltman CJ (1988) Song sharing in a group-living songbird, the Australian magpie, *Gymnorhina tibicen*, I. Vocal sharing within and amongst social groups. *Behaviour* **104**(1), 1–27. doi:10.1163/156853988X00575

Brown ED, Farabaugh SM, Hughes JM (1993) A test of centre-edge hypotheses in a permanently territorial songbird, the Australian magpie, *Gymnorhina tibicen*. *Animal Behaviour* **45**(4), 814–816. doi:10.1006/anbe.1993.1095

Brown JL (1987) *Helping and Communal Breeding in Birds*. Princeton University Press, Princeton.

Bugnyar T, Kotrschal K (2002) Raiding food caches in ravens, *Corvus corax*. *Animal Behaviour* **64**(2), 185–195. doi:10.1006/anbe.2002.3056

Buitron D, Nuechterlein GL (1985) Experiments on olfactory detection of food caches by black-billed magpies. *The Condor* **87**, 92–95. doi:10.2307/1367139

Burne THJ, Rogers LJ (1999) Changes in olfactory responsiveness by the domestic chick after early exposure to odorants. *Animal Behaviour* **58**, 329–336. doi:10.1006/anbe.1999.1151

Busnell RG (1977) Acoustic communication. In *How Animals Communicate*. (Ed. TA Seboek) pp. 233–252. Indiana University Press, Bloomington.

Cake M, Black A, Joseph L (2018) The generic taxonomy of the Australian Magpie and Australo-Papuan butcherbirds is not all black-and-white. *Bulletin of the British Ornithologists' Club* **138**(4), 346–359. doi:10.25226/bboc.v138i4.2018.a6

Campbell AJ (1895) *G. dorsalis*, n. sp. (The Long-billed Magpie). *Proceedings of the Royal Society of Victoria* **7**(2), 206–208.

Campbell AG (1929) Australian magpies of the genus *Gymnorhina*. *Emu* **28**(3), 165–176.

Carrick R (1963) Ecological significance of territories in the Australian Magpie. *Proceedings of the International Ornithological Congress* **13**, 740–753.

Carrick R (1972) Population ecology of the Australian black-backed magpie, royal penguin and silver gull. In *Population Ecology of Migratory Birds: A symposium*. US Department of the Interior. Wildlife Research Report **2**, 41–99.

Carrick R (1984) Case study of the Australian magpie. In *The Ecological Web: More on the Distribution and Abundance of Animals*. (Eds HG Andrewartha and LC Birch) pp. 372–387. The University of Chicago Press, Chicago.

Catry P, Furness RW (1997) Territorial intrusions and copulation behaviour in the great skua. *Animal Behaviour* **54**, 1265–1272. doi:10.1006/anbe.1997.0543

Catterall CP, Green RJ, Jones DN (1991) Habitat use by birds across a forest–suburb interface in Brisbane: implications for corridors. In *Nature Conservation 2: The Role of Corridors*. (Eds DA Saunders and RJ Hobb) pp. 247–258. Surrey Beatty & Sons, Sydney.

Ceballos G, Ehrlich PE, Barnosky AD, García A, Pringle RM, Palmer TM (2015) Accelerated modern human–induced species losses: entering the sixth mass extinction. *Science Advances* **1**, e1400253.

Chapman G (1978) Caching of food by the Australian crow. *Emu* **78**, 98. doi:10.1071/MU9780098b

Chen Y, Matheson LE, Sakata JT (2016) Mechanisms underlying the social enhancement of vocal learning in songbirds. *Proceedings of the National Academy of Sciences of the United States of America* **113**(24), 6641–6646.

Chisholm AH (1948) *Bird Wonders of Australia*. Angus & Robertson, Sydney.

Christensen C, Radford AN (2018) Dear enemies or nasty neighbors? Causes and consequences of variation in the responses of group-living species to territorial intrusions. *Behavioral Ecology* **29**(5), 1004–1013. doi:10.1093/beheco/ary010

Christidis L, Boles WE (1994) *The Taxonomy and Species of Birds of Australia and Its Territories*. Monograph, Royal Australasian Ornithologists Union, Hawthorn East, Vic.

Christidis L, Boles WE (2008) *Systematics and Taxonomy of Australian Birds*. p. 174. CSIRO Publishing, Melbourne.

Christidis L, Norman JA (2010) Evolution of the Australasian songbird fauna. *Emu* **110**, 21–31. doi:10.1071/MU09031

Clary D, Cheys A, Kelly DM (2014) Pattern of visuospatial lateralization in two corvid species, black-billed magpies and Clark's nutcrackers. *Behavioural Processes* **107**, 94–98. doi:10.1016/j.beproc.2014.07.020

Clay T (1950) The Mallophaga as an aid to the classification of birds with special reference to the structure of feathers. *Proceedings. Xth International Ornithological Congress*, pp. 207–215.

Clayton NS, Dickinson A (1998) Episodic-like memory during cache recovery by scrub jays. *Nature* **395**, 272–274. doi:10.1038/26216

Clayton NS, Dickinson A (1999a) Scrub jays (*Aphelocoma coerulescens*) remember the relative time of caching as well as the location and content of their caches. *Journal of Comparative Psychology* **113**, 403–416. doi:10.1037/0735-7036.113.4.403

Clayton NS, Dickinson A (1999b) Memory for the contents of caches by scrub jays. *Journal of Experimental Psychology. Animal Behavior Processes* **25**, 82–91. doi:10.1037/0097-7403.25.1.82

Clayton NS, Soha JA (1999) Memory in avian food caching and song learning: a general mechanism or different processes? *Advances in the Study of Behavior* **28**, 115–173. doi:10.1016/S0065-3454(08)60217-X

Clayton NS, Dally J, Gilbert J, Dickinson A (2005) Food caching by western scrub-jays (*Aphelocoma californica*) is sensitive to the conditions at recovery. *Journal of Experimental Psychology. Animal Behavior Processes* **31**(2), 115. doi:10.1037/0097-7403.31.2.115

Cockburn A (1996) Why do so many Australian birds cooperate: social evolution in the Corvidae. In *Frontiers in Population Ecology*. (Eds RB Floyd, AW Sheppard and PJ de Barro) pp. 451–472. CSIRO Publishing, Melbourne.

Cockburn A (1998) Evolution of helping behavior in cooperatively breeding birds. *Annual Review of Ecology and Systematics* **29**(1), 141–177.

Cockburn A (2006) Prevalence of different modes of parental care in birds. *Proceedings. Biological Sciences* **273**, 1375–1383 doi:10.1098/rspb.2005.3458

Collins J (1983) Magpie mimic. *Geelong Naturalist* **20**(3), 80.

Cooney R, Cockburn A (1995) Territorial defense is the major function of female song in the superb fairy-wren, *Malurus cyaneus*. *Animal Behaviour* **49**, 1635–1647.

Cooper A, Penny D (1997) Mass survival of birds across the Cretaceous-Tertiary boundary: molecular evidence. *Science* **275**(5303), 1109–1113. doi:10.1126/science.275.5303.1109

Corballis MC (2012) Lateralization of the human brain. *Progress in Brain Research* **195**, 103–121. doi:10.1016/B978-0-444-53860-4.00006-4

Corfield JR, Long B, Krilow JM, Wylie DR, Iwaniuk AN (2016) A unique cellular scaling rule in the avian auditory system. *Brain Structure & Function* **221**, 2675–2693. doi:10.1007/s00429-015-1064-1

Courvoisier H, Camacho-Schlenker S, Aubin T (2014) When neighbours are not 'dear enemies': a study in the winter wren, *Troglodytes troglodytes*. *Animal Behaviour* **90**, 229–235.

Cracraft J (2001) Avian evolution, Gondwana biogeography and the Cretaceous-Tertiary mass extinction event. *Proceedings. Biological Sciences* **268**, 459–469. doi:10.1098/rspb.2000.1368

Cracraft J, Barker FK, Braun M, Harshman J, Dyke GJ, Feinstein J, Stanley S, Cibois A, Schikler P, Beresford P, García-Moreno J, Sorenson MD, Yuri T, Mindell DP (2004) Phylogenetic relationships among modern birds (Neornithes) – toward an avian tree of life. In *Assembling the Tree of Life*. (Eds J Cracraft and MJ Donoghue) pp. 468–489. Oxford University Press, New York.

Cresswell ID, Murphy HT (2017) *Australia State of the Environment 2016: Biodiversity*. Independent report to the Australian Government Minister for the Environment and Energy. Australian Government Department of the Environment and Energy, Canberra, ACT.

Cruickshank AJ, Gautier J-P, Chappuis C (1993) Vocal mimicry in wild African Grey Parrots *Psittacus erithacus*. *Ibis* **135**, 293–299.

Curio E (1988) Cultural transmission of enemy recognition by birds. In *Social Learning: Psychological and Biological Perspective*. (Eds TR Zentall and BG Galef, Jr.) pp. 75–97. Lawrence Erlbaum, Hillsdale, NJ.

D'Antonio-Bertagnolli A, Anderson MJ (2018) Lateral asymmetry in the freely occurring behaviour of budgerigars (*Melopsittacus undulatus*) and its relation to cognitive performance. *Laterality* **23**(3), 344–363. doi:10.1080/1357650X.2017.1361964

Darwin C (1871) *The Descent of Man and Selection in Relation to Sex*. John Murray, London.

Darwin C (1904) *The Expression of the Emotions in Man and Animals*. John Murray, London.

Davidson S (2002) Facing extinction. *Ecos* **113**, 24–30.

Davis RA, Wilcox JA (2013) Adapting to suburbia: bird ecology on an urban-bushland interface in Perth, Western Australia. *Pacific Conservation Biology* **19**(2), 110–120.

Dawkins R (1976) *The Selfish Gene*, Oxford University Press, Oxford.

de Boer HJ, Steffen K, Cooper WE (2015) Sunda to Sahul dispersals in Trichosanthes (Cucurbitaceae): a dated phylogeny reveals five independent dispersal events to Australasia. *Journal of Biogeography* **42**, 519–531. doi:10.1111/jbi.12432

De Kort SR, Tebbich S, Dally JM, Emery NJ, Clayton NS (2006) The comparative cognition of caching. In *Comparative Cognition: Experimental Explorations of Animal*

Intelligence. (Eds EA Wasserman and TR Zentall) pp. 529–552. Oxford University Press, New York.

De Waal FB (2000) Attitudinal reciprocity in food sharing among brown capuchin monkeys. *Animal Behaviour* **60**(2), 253–261.

Debus SJS (1996) Magpies, currawongs and butcherbirds. In *Finches, Bowerbirds and other Passerines of Australia*. (Ed. R Strahan) pp. 236–262. Angus & Robertson, Sydney.

Deng C, Kaplan G, Rogers LJ (2001) Similarity of the song control nuclei of male and female Australian magpies (*Gymnorhina tibicen*). *Behavioural Brain Research* **123**(1), 89–102. doi:10.1016/S0166-4328(01)00200-5

Derryberry EP (2009) Ecology shapes birdsong evolution: variation in morphology and habitat explains variation in white-crowned sparrow song. *American Naturalist* **174**(1), 24–33.

Descartes R (2004) [1637] *A Discourse on Method, Meditations and Principles*. Trans. J Veitch. Orion Publishing Group, London.

Dickinson EC (2003) *The Howard and Moore Complete Checklist of the Birds of the World*. Christopher Helm, London.

Dickinson EC, Christidis L (Eds) (2014) *The Howard and Moore Complete Checklist of the Birds of the World, Vol. 2: Passerines*. 4th edn. pp. 205–208. Aves Press, Eastbourne, UK.

Double MC, Cockburn A (2000) Pre-dawn infidelity: females control extra-pair mating in superb fairy-wrens. *Proceedings. Biological Sciences* **267**, 465–470. doi:10.1098/rspb.2000.1023

Dovers S (1994) Australian environmental history: introduction, review and principles. In *Australian Environmental History. Essays and Cases*. (Ed. S Dovers) pp. 2–19. Oxford University Press, Melbourne.

Downing PA, Cornwallis CK, Griffin AS (2017) How to make a sterile helper. *BioEssays* **39**(1), 1600136. doi:10.1002/bies.201600136

Dowsett-Lemaire F (1979) The imitative range of the song of the marsh warbler *Acrocephalus palustris*, with special reference to imitations of African birds. *Ibis* **121**, 453–468.

Drayson N (2002) The Magpie: Australia's iconic bird. *Australian Geographic* **68**, 36–47.

Drinkwater E, Ryeland J, Haff T, Umbers KDL (2017) A novel observation of food dunking in the Australian Magpie *Gymnorhina tibicen*. *Australian Field Ornithology* **34**, 95–97. doi:10.20938/afo34095097

Duecker F, Formisano E, Sack AT (2013) Hemispheric differences in the voluntary control of spatial attention: direct evidence for a right-hemispheric dominance within frontal cortex. *Journal of Cognitive Neuroscience* **25**, 1332–1342. doi:10.1162/jocn_a_00402

Dugatkin LA (1997) *Cooperation Among Animals: An Evolutionary Perspective*. Oxford University Press, New York.

Duncan RA (2002) A time frame for construction of the Kerguelen Plateau and Broken Ridge. *Journal of Petrology* **43**, 1109–1119. doi:10.1093/petrology/43.7.1109

Dunham ML, Warner RR, Lawson JW (1995) The dynamics of territory acquisition: a model of two coexisting strategies. *Theoretical Population Biology* **47**, 347–364. doi:10.1006/tpbi.1995.1016

Dunn PO, Cockburn A (1999) Extrapair mate choice and honest signalling in cooperatively breeding superb fairy-wrens. *Evolution* **53**, 938–946. doi:10.1111/j.1558-5646.1999.tb05387.x

Durrant KL, Hughes JM (2005) Differing rates of extra-group paternity between two populations of the Australian magpie (*Gymnorhina tibicen*). *Behavioral Ecology and Sociobiology* **57**, 536–545. doi:10.1007/s00265-004-0883-5

Eales LA (1989) The influences of visual and vocal interactions on song learning in zebra finches. *Animal Behaviour* **37**, 507–508. doi:10.1016/0003-3472(89)90097-3

East R, Pottinger RP (1975) Starling (*Sturnus vulgaris* L.) predation on grass grub (*Costelytra zealandica* (White), Melolonthinae) populations in Canterbury. *New Zealand Journal of Agricultural Research* **18**, 417–452. doi:10.1080/00288233.1975.10421071

Edwards SV, Boles WE (2002) Out of Gondwana: the origin of passerine birds. *Trends in Ecology & Evolution* **17**(8), 347–349.

Edwards EK, Mitchell NJ, Amanda R, Ridley AR (2015) The impact of high temperatures on foraging behaviour and body condition in the Western Australian Magpie *Cracticus tibicen dorsalis*. *Ostrich* **86**(1–2), 137–144. doi:10.2989/00306525.2015.1034219

Emery NJ, Clayton NS (2004) Comparing the complex cognition of birds and primates. In *Comparative Vertebrate Cognition: Are Primates Superior to Non-primates.* (Eds LJ Rogers and G Kaplan) pp. 3–55. Kluwer Academic/Plenum Publishers, New York.

Emlen S (1972) An experimental analysis of the parameters of bird song eliciting species recognition. *Behaviour* **41**, 130–171. doi:10.1163/156853972X00248

Evans CS (1997) Referential signals. *Perspectives in Ethology* **12**, 99–143. doi:10.1007/978-1-4899-1745-4_5

Evans CS, Evans L (1999) Chicken food calls are functionally referential. *Animal Behaviour* **58**, 307–319. doi:10.1006/anbe.1999.1143

Evans CS, Evans L, Marler P (1993) On the meaning of alarm calls: functional reference in an avian vocal system. *Animal Behaviour* **46**, 23–38. doi:10.1006/anbe.1993.1158

Falls J, Brooks R (1975) Individual recognition by song in white-throated sparrows. II. Effects of location. *Canadian Journal of Zoology* **53**, 1412–1420. doi:10.1139/z75-170

Farabaugh SM (1982) The ecological and social significance of duetting. In *Acoustic Communication in Birds (Vol. 2).* (Eds DE Kroodsma and EH Miller) pp. 85–124. Academic Press, New York.

Farabaugh SM, Brown ED, Veltman CJ (1988) Song sharing in a group-living songbird, the Australian magpie, *Gymnorhina tibicen*, II. Vocal sharing between territorial neighbors, within and between geographic regions, and between sexes. *Behaviour* **104**(1), 105–125. doi:10.1163/156853988X00629

Farabaugh SM, Brown ED, Hughes JM (1992) Cooperative territorial defence in the Australian magpie, *Gymnorhina tibicen* (Passeriformes, Cracticidae), a group-living songbird. *Ethology* **92**(4), 283–292. doi:10.1111/j.1439-0310.1992.tb00966.x

Farabaugh SM, Linzenbold A, Dooling RJ (1994) Vocal plasticity in budgerigars (*Melopsittacus undulatus*): Evidence for social factors in the learning of contact calls. *Journal of Comparative Psychology* **108**(1), 81–92. doi:10.1037/0735-7036.108.1.81

Farrow-Smith E, Marciniak C, Shoebridge J (2018) Magpie swooping a playful game of cat and mouse between boy and feathered friend. 12 Sept. ABC North Coast. <https://www.abc.net.au/news/2018-09-12/the-boy-the-magpie-and-the-daily-dash/10235662>

Feduccia AA, Duvauchelle CL (2008) Auditory stimuli enhance MDMA-conditioned reward and MDMA-induced nucleus accumbens dopamine, serotonin and locomotor responses. *Brain Research Bulletin* **77**, 189–196. doi:10.1016/j.brainresbull.2008.07.007

Feeney WE, Medina I, Somveille M, Heinsohn R, Hall ML, Mulder RA, Stein JA, Kilner RM, Langmore NE (2013) Brood parasitism and the evolution of cooperative breeding in birds. *Science* **342**(6165), 1506–1508. doi:10.1126/science.1240039

Ficken MS (1977) Avian play. *The Auk* **94**, 573–582.

Ficken MS (1990) Acoustic characteristics of alarm calls associated with predation risk in chickadees. *Animal Behaviour* **39**, 400–401. doi:10.1016/S0003-3472(05)80888-7

Finn P, Hughes JM (1995) Relationship of helpers to offspring in the Australian magpie *Gymnorhina tibicen*. *Australian Society for the Study of Animal Behaviour, 22nd Annual Conference, 20–22 April 1995*, Brisbane, Griffith University.

Finn PG, Hughes JM (2001) Helping behaviour in Australian magpies, *Gymnorhina tibicen*. *Emu* **101**, 1–7.

Fischer B, Van Doorn GS, Dieckmann U, Taborsky B (2014) The evolution of age-dependent plasticity. *American Naturalist* **183**, 108–125. doi:10.1086/674008

Fisher J (1954) Evolution and bird sociality. In *Evolution as a process*. (Eds JA Huxley, C Hardy, EB Ford) pp. 71–83. Allen & Unwin, London.

Floyd RB, Woodland DJ (1981) Localization of soil dwelling scarab larvae by the black-backed magpie, *Gymnorhina tibicen* (Latham). *Animal Behaviour* **29**, 510–517. doi:10.1016/S0003-3472(81)80112-1

Ford HA (1989) *Ecology of Birds: An Australian Perspective*. Surrey Beatty & Sons, Chipping Norton, New South Wales.

Fortune ES, Margoliash D (1995) Parallel pathways and convergence onto HVC and adjacent neostriatum of adult zebra finches (*Taeniopygia guttata*). *The Journal of Comparative Neurology* **360**, 413–441. doi:10.1002/cne.903600305

Franklin DC, Garnett S, Luck G, Gutierrez-Ibanez C, Iwaniuk A (2014) Relative brain size in Australian birds. *Emu* **114**, 160–170.

Frey FA, Coffin MF, Wallace PJ, Weis D (Eds) (2003) Leg 183 synthesis: Kerguelen Plateau–Broken Ridge – a large igneous province. *Proceedings of the Ocean Drilling Program, Scientific Results* **183**, 1–48.

Frith HJ (Ed.) (1969) *Birds in the Australian High Country*. AH & AW Reed, Sydney.

Fuchs J, Irestedt M, Fieldså J, Couloux A, Pasquet E, Bowie RCK (2012) Molecular phylogeny of African bush-shrikes and allies: tracing the biogeographic history of an explosive radiation of corvoid birds. *Molecular Phylogenetics and Evolution* **64**, 93–105. doi:10.16/j.ympev.2012.03.007

Fulton GR (2006) Plural-breeding Australian Magpies *Gymnorhina tibicen dorsalis* nesting annually in the same tree. *Australian Field Ornithology* **23**, 198–201.

Galván I (2008) The importance of white on black: unmelanized plumage proportion predicts display complexity in birds. *Behavioral Ecology and Sociobiology* **63**(2), 303. doi:10.1007/s00265-008-0662-9

Gardner T, Gardner P (1975) Observations at a magpie's nest. *Bird Watcher* **6**(3), 82–84.

Gaunt AS, Gaunt SLL (1985) Syringeal structure and avian phonation. In *Current Ornithology, Vol. II*. (Ed. RF Johnston) pp. 213–245. Plenum Press, New York.

Gibbs H (2007) Climatic variation and breeding in the Australian magpie (*Gymnorhina tibicen*): a case study using existing data. *Emu* **107**, 284–293. doi:10.1071/MU07022

Gill F, Donsker D (Eds) (2018) IOC World Bird List (v.8.2). doi:10.14344/IOC.ML.8.2

Gill FB, Donsker D (Eds) (2017) Bristlehead, Butcherbirds, Woodswallows & Cuckooshrikes. *World Bird List Version 7.3*. International Ornithologists' Union.

Giraldeau LA, Ydenberg R (1987) The centre-edge effect: the result of a war of attrition between territorial constraints. *The Auk* **104**, 535–538. doi:10.2307/4087559

Gobes SM, Bolhuis JJ (2007) Birdsong memory: a neural dissociation between song recognition and production. *Current Biology* **17**, 789–793. doi:10.1016/j.cub.2007.03.059

Goller F, Larsen ON (1997) A new mechanism of sound generation in songbirds. *Proceedings of the National Academy of Sciences of the United States of America* **80**, 2390–2394.

Goodyer GJ, Nicholas A (2007) Scarab grubs in northern tableland pastures. *Primefact* **512**, 1–8.

Gould J (1837) *A Synopsis of the Birds of Australia, and the Adjacent Islands; 1837–38*. Published by the author.

Green JP, Freckleton RP, Hatchwell BJ (2016) Variation in helper effort among cooperatively breeding bird species is consistent with Hamilton's Rule. *Nature Communications* **7**, 12663. doi:10.1038/ncomms12663

Griffioen PA, Clarke MF (2002) Large-scale bird-movement patterns evident in eastern Australian atlas data. *Emu* **102**, 99–125. doi:10.1071/MU01024

Guillette LM, Scott AC, Healy SD (2016) Social learning in nest-building birds: a role for familiarity. *Proceedings of the Royal Society B* **283**(1827), 20152685.

Güntürkün O, Bugnyar T (2016) Cognition without cortex. *Trends in Cognitive Sciences* **20**(4), 291–303. doi:10.1016/j.tics.2016.02.001

Güntürkün O, Ocklenburg S (2017) Ontogenesis of lateralization. *Neuron* **94**(2), 249–263. doi:10.1016/j.neuron.2017.02.045

Habib R, Nyberg L, Tulving E (2003) Hemispheric asymmetries of memory: the HERA model revisited. *Trends in Cognitive Sciences* **7**, 241–245. doi:10.1016/S1364-6613(03)00110-4

Haftorn S (2000) Context and possible functions of alarm calling in the willow tit, *Parus montanus*. The principle of 'better safe than sorry'. *Behaviour* **137**, 437–449.

Hailman JP (1998) The avian gait. *Abstracts. 22nd International Ornithological Congress*. Durban, South Africa.

Hancock P (2013) Magpies have chortled since the very first dawn. *Sydney Morning Herald*, 8 March. <https://www.smh.com.au/entertainment/magpies-have-chortled-since-the-very-first-dawn-20130307-2fmo5.html>.

Harris LJ (1989) Footedness in parrots: three centuries of research, theory, and mere surmise. *Canadian Journal of Psychology* **43**, 369–396. doi:10.1037/h0084228

Harshman J (2006) Oscines. Songbirds. Version 31 July 2006 (under construction). http://tolweb.org/Oscines/29222/2006.07.31 in The Tree of Life Web Project, http://tolweb.org/

Hartley RJ, Suthers RA (1990) Lateralisation of syringeal function during song production in the canary. *Journal of Neurobiology* **21**, 1236–1248. doi:10.1002/neu.480210808

Hausberger M (1997) Social influences on song acquisition and sharing in the European starling (*Sturnus vulgaris*). In *Social Influences on Vocal Development*. (Eds CT Snowdon and M Hausberger) pp. 128–156. Cambridge University Press, Cambridge, UK.

Hausberger M, Jenkins PF, Keene J (1991) Species-specificity and mimicry in bird song: are they paradoxes? A re-evaluation of song mimicry in the European starling. *Behaviour* **117**, 53–81. doi:10.1163/156853991X00120

Heather BD, Robertson HA (1996) *Field Guide to the Birds of New Zealand*. Viking, Auckland.

Heinrich B, Pepper JW (1998) Influence of competitors on caching behavior in the common raven. *Animal Behaviour* **56**, 1083–1090. doi:10.1006/anbe.1998.0906

Heinrich B, Smolker R (1998) Play in common ravens (*Corvus corax*). In *Animal Play: Evolutionary, Comparative and Ecological Perspectives*. (Eds M Bekoff and JA Byers) pp. 27–44. Cambridge University Press, Cambridge, UK.

Heinsohn R, Cockburn A (1994) Helping is costly to young birds in cooperatively breeding white-winged choughs. *Proceedings. Biological Sciences* **256**, 293–298. doi:10.1098/rspb.1994.0083

Heinsohn RG, Double MC (2004) Cooperate or speciate: new theory for the distribution of passerine birds. *Trends in Ecology & Evolution* **19**(2), 55–57. doi:10.1016/j.tree.2003.12.001

Helgesen IM, Hamblin S, Hurd PL (2013) Does cheating pay? Re-examining the evolution of deception in a conventional signalling game. *Animal Behaviour* **86**, 1215–1224. doi:10.1016/j.anbehav.2013.09.023

Henderson J, Hurly TA, Bateson M, Healy SD (2006) Timing in free-living rufous hummingbirds, *Selasphorus rufous*. *Current Biology* **16**, 512–515. doi:10.1016/j.cub.2006.01.054

Heppner F (1965) Sensory mechanisms and environmental cues used by the American robin in locating earthworms. *The Condor* **67**, 247–256. doi:10.2307/1365403

Herculano-Houzel S, Lent R (2005) Isotropic fractionator: a simple, rapid method for the quantification of total cell and neuron numbers in the brain. *The Journal of Neuroscience* **25**, 2518–2521. doi:10.1523/JNEUROSCI.4526-04.2005

Hinde RA (1956) The biological significance of territories in birds. *Ibis* **98**, 340–369.

Hinde RA (Ed.) (1969) *Bird Vocalisations*. Cambridge University Press, Cambridge, UK.

Hinde RA (Ed.) (1972) Non-verbal communication. Cambridge University Press, Cambridge, UK.

Hindmarsh AM (1984) Vocal mimicry in starlings. *Behaviour* **90**, 302–325.

Hobbs JN (1971) Use of tools by the white-winged chough. *Emu* **71**, 84–85. doi:10.1071/MU971084a

Hoffman AM, Robakiewicz PE, Tuttle EM, Rogers LJ (2006) Behavioural lateralisation in the Australian magpie (*Gymnorhina tibicen*). *Laterality* **11**, 110–121. doi:10.1080/13576500500376674

Hohtola E, Visser GH (1998) Development of locomotion and endothermy in altricial and precocial birds. In *Avian Growth and Development*. (Eds RE Ricklefs and JM Starck) pp. 157–173. Oxford University Press, New York.

Hollén LI, Radford AN (2009) The development of alarm call behaviour in mammals and birds. *Animal Behaviour* **78**, 791–800. doi:10.1016/j.anbehav.2009.07.021

Howe RW (1984) Local dynamics of bird assemblages in small forest habitat islands in Australia and North America. *Ecology* **65**, 1585–1601. doi:10.2307/1939138

Huber L, Rechberger S, Taborsky M (2001) Social learning affects object exploration and manipulation in keas, *Nestor notabilis*. *Animal Behaviour* **62**, 945–954. doi:10.1006/anbe.2001.1822

Hughes JM (1980) Geographic variation in the Australian magpie and its mallophagan parasites. PhD thesis, Zoology Department, La Trobe University, Melbourne.

Hughes JM (1983) An explanation for the asymmetrical 'hybrid' zone between black-backed and white-backed magpies. *Emu* **82**, 50–53. doi:10.1071/MU9820050

Hughes JM (1984a) Distribution of mallophaga on the Australian magpie (family Cracticidae). *Australian Journal of Zoology* **21**(5), 459–466.

Hughes JM (1984b) Morphometric variation in the mallophaga of the Australian magpie (*Gymnorhina tibicen*, Latham). *Australian Journal of Zoology* **21**(5), 467–477.

Hughes JM, Mather PB (1991) Variation in the size of territorial groups in the Australian magpie *Gymnorhina tibicen*. *Proceedings of the Royal Society of Queensland* **101**(30 January), 13–19.

Hughes JM, Pearce BJ, Vockenson K (1983) Territories of the Australian magpie *Gymnorhina tibicen* in South-East Queensland. *Emu* **83**, 108–111. doi:10.1071/MU9830108

Hughes JM, Hesp JDE, Kallioinen R, Kempster M, Lange CL, Hedstrom KE, Mather PB, Robinson A, Wellbourn MJ (1996) Differences in social behaviour between populations of the Australian magpie *Gymnorhina tibicen*. *Emu* **96**, 65–70. doi:10.1071/MU9960065

Hughes JM, Mather PB, Toon A, Rowley I, Russell E (2003) High levels of extragroup paternity in a population of Australian magpies *Gymnorhina tibicen*: evidence from microsatellites. *Molecular Ecology* **12**, 3441–3450. doi:10.1046/j.1365-294X.2003.01997.x

Hultsch H, Todt D (1989) Memorization and reproduction of songs in nightingales: evidence for package formation. *Journal of Comparative Physiology. A, Neuroethology, Sensory, Neural, and Behavioral Physiology* **165**, 197–203. doi:10.1007/BF00619194

Hunt GR, Rutledge RB, Gray RD (2006) The right tool for the job: what strategy do wild New Caledonian crows use? *Animal Cognition* **9**, 307–316. doi:10.1007/s10071-006-0047-2

Hurley S, Nudds M (Eds.) (2006) *Rational Animals?* Oxford University Press, New York.

Hutchinson JM, Gigerenzer G (2005) Simple heuristics and rules of thumb: Where psychologists and behavioural biologists might meet. *Behavioural Processes* **69**(2), 97–124. doi:10.1016/j.beproc.2005.02.019

Hyman J (2005) Seasonal variation in response to neighbours and strangers by a territorial songbird. *Ethology* **111**, 951–961. doi:10.1111/j.1439-0310.2005.01104.x

Igic B, McLachlan J, Lehtinen I, Magrath RD (2015) Crying wolf to a predator: deceptive vocal mimicry by a bird protecting young. *Proceedings of the Royal Society B* **282**(1809), 20150798.

Ihle M, Kempenaers B, Forstmeier W (2015) Fitness benefits of mate choice for compatibility in a socially monogamous species. *PLoS Biology* **13**(9), e1002248. doi:10.1371/journal.pbio.1002248

Ishigame G, Baxter GS, Lisle AT (2006) Effects of artificial foods on the blood chemistry of the Australian magpie. *Austral Ecology* **31**, 199–207. doi:10.1111/j.1442-9993.2006.01580.x

Itakura S (2004) Gaze-following and joint visual attention in nonhuman animals. *The Japanese Psychological Research* **46**(3), 216–226.

Iwaniuk AN, Heesy CP, Hall MI, Wylie DR (2008) Relative Wulst volume is correlated with orbit orientation and binocular visual field in birds. *Journal of Comparative Physiology. A, Neuroethology, Sensory, Neural, and Behavioral Physiology* **194**, 267–282. doi:10.1007/s00359-007-0304-0

Izawa E-I, Kusayama T, Watanabe S (2005) Foot-use laterality in the Japanese jungle crow (*Corvus macrorhynchos*). *Behavioural Processes* **69**, 357–362. doi:10.1016/j.beproc.2005.02.001

Jackson WJ, Argent RM, Bax NJ, Bui E, Clark GF, Coleman S, Cresswell ID, Emmerson KM, Evans K, Hibberd MF, Johnston EL, Keywood MD, Klekociuk A, Mackay R, Metcalfe D, Murphy H, Rankin A, Smith DC, Wienecke B (2016) Overview. In *Australia State of the Environment 2016*. Australian Government Department of the Environment and Energy, Canberra, <https://soe.environment.gov.au/theme/overview>.

Jacquin L, Lenouvel P, Haussy C, Ducatez S, Gasparini J (2011) Melanin-based coloration is related to parasite intensity and cellular immune response in an urban free-living bird: the feral pigeon *Columba livia*. *Journal of Avian Biology* **42**(1), 11–15.

Janik VM (2014) Cetacean vocal learning and communication. *Current Opinion in Neurobiology* **28**, 60–65. doi:10.1016/j.conb.2014.06.010

Jarvis ED, Güntürkün O, Bruce L, Csillag A, Karten H, Kuenzel W, Medina L, Paxinos G, Perkel DJ, Shimizu T, Striedter G (2005) Avian brains and a new understanding of vertebrate brain evolution. *Nature Reviews. Neuroscience* **6**(2), 151. doi:10.1038/nrn1606

Jelbert SA, Taylor AH, Gray RD (2016) Does absolute brain size really predict self-control? Hand-tracking training improves performance on the A-not-B task. *Biology Letters* **12**, 20150871. doi:10.1098/rsbl.2015.0871

Jenkins PF (1978) Cultural transmission of song patterns and dialect development in a free-living bird population. *Animal Behaviour* **26**, 50–78. doi:10.1016/0003-3472(78)90007-6

Johnson G, Jones D (1998) Whistling in the dark: 'Pre-dawn song', a 'new' model of vocal communication in birds. *ASSAB Conference, Abstracts*, p. 17.

Johnson LS, Kermott LH (1990) Possible causes of territory takeovers in a north temperate population of house wrens. *The Auk* **107**, 781–784. doi:10.2307/4088013

Johnstone RA (2011) Load lightening and negotiation over offspring care in cooperative breeders. *Behavioral Ecology* **22**, 436–444. doi:10.1093/beheco/arq190

Jones AE, Ten Cate C, Slater PJ (1996) Early experience and plasticity of song in adult male zebra finches (*Taeniopygia guttata*). *Journal of Comparative Psychology* **110**(4), 354.

Jones DN (1996) Magpie Attack! Our best-known, least-understood wildlife problem. *Wingspan* **6**, 12–15.

Jones DN (2002) *Magpie Alert: Learning to Live with a Wild Neighbour*. UNSW Press, Sydney.

Jones DN, Finn PG (1999) Translocation of aggressive Australian magpies: a preliminary assessment of a potential management action. *Wildlife Research* **26**, 271–279. doi:10.1071/WR98062

Jones DN, Thomas LK (1998) Managing to live with Brisbane's wildlife: magpies and the management of positive and negative interactions. *Proceedings of the Royal Society of Queensland* **107**, 45–49.

Jones DN, Beard N, Brooks J, Heard R, Nealson T, Rollinson D, Warne R (1999) 'Magpie aggression toward humans: Results and assessment of 1999 magpie season'. Report to Queensland Parks and Wildlife Service, Queensland Government, Brisbane.

Jønsson KA, Fabre P-H, Ricklefs RE, Fjeldså J (2011) Major global radiation of corvoid birds originated in the proto-Papuan archipelago. *Proceedings of the National Academy of Sciences of the United States of America* **108**, 2328–2333. doi:10.1073/pnas.1018956108

Jønsson KA, Fabre P-H, Irestedt M (2012) Brains, tools, innovation and biogeography in crows and ravens. *BMC Evolutionary Biology* **12**, 72. doi:10.1186/1471-2148-12-72

Jønsson KA, Fabre P-H, Kennedy JD, Holt BG, Borregaard MK, Rahbek C, Fjeldså J (2016) A supermatrix phylogeny of corvoid passerine birds (Aves: Corvides). *Molecular Phylogenetics and Evolution* **94A**, 87–94. doi:10.1016/j.ympev.2015.08.020

Jurisevic MA (1999) Structural change of begging vocalisations and vocal repertoires in two hand-raised Australian passerines, the little raven *Corvus mellori* and whitewinged chough *Corcorax melanorhamphos*. *Emu* **99**, 1–8. doi:10.1071/MU99001

Jurisevic MA (2003) Convergent characteristics of begging vocalisations in Australian birds. *Lundiana* **4**(1), 25–33.

Jurisevic MA, Sanderson KJ (1994) Alarm vocalisations in Australian birds: convergent characteristics and phylogenetic differences. *Emu* **94**, 69–77. doi:10.1071/MU9940067

Jurisevic MA, Sanderson KJ (1998a) A comparative analysis of distress call structure in Australian passerine and non-passerine species: influence of size and phylogeny. *Journal of Avian Biology* **29**, 61–71. doi:10.2307/3677342

Jurisevic MA, Sanderson KJ (1998b) Acoustic discrimination of passerine antipredator signals by Australian raptors. *Australian Journal of Zoology* **46**, 369–379. doi:10.1071/ZO97052

Kabadayi C, Taylor LA, von Bayern AMP, Osvath M (2016) Ravens, New Caledonian crows and jackdaws parallel great apes in motor self-regulation despite smaller brains. *Royal Society Open Science* **3**, 160104. doi:10.1098/rsos.160104

Kallioinen RUO, Hughes JM, Mather PB (1995) Significance of back colour in territorial interactions in the Australian magpie. *Australian Journal of Zoology* **43**, 665–673. doi:10.1071/ZO9950665

Kaplan G (1996) Selective learning and retention: song development and mimicry in the Australian magpie. *International Journal of Psychology* **31**, 233.

Kaplan G (1998) Vocal mimicry and territoriality in the Australian Magpie *Gymnorhina tibicen*, *Ostrich* **69**, 256.

Kaplan G (2000) Song structure and function of mimicry in the Australian magpie (*Gymnorhina tibicen*) compared to the lyrebird (*Menura* ssp.). *International Journal of Comparative Psychology* **12**(4), 219–241.

Kaplan G (2003) Magpie mimicry. *Nature Australia* **27**(10), 60–67.

Kaplan G (2004) Magpie mimicry. In *Encyclopedia of Animal Behaviour, vol. 2*. (Eds M Bekoff and J Goodall) pp. 772–774. Greenwood Publishing, Westport, CT.

Kaplan G (2005) The vocal behaviour of Australian Magpies (*Gymnorhina tibicen*): a study of vocal development, song learning, communication and mimicry in the Australian Magpie. PhD thesis, School of Veterinary Sciences, University of Queensland, St Lucia, Brisbane.

Kaplan G (2006a) Australian magpie voice. In *Handbook of Australian, New Zealand and Antarctic Birds, Vol. 7(A)*. (Ed. J Peter) pp. 605–608, 613–616. Oxford University Press, Melbourne.

Kaplan G (2006b) Alarm calls, communication and cognition in Australian magpies (*Gymnorhina tibicen*). *Acta Zoologica Sinica* **52**(Supplement), 614–617.

Kaplan G (2008a) Alarm calls and referentiality in Australian magpies: between midbrain and forebrain, can a case be made for complex cognition? *Brain Research Bulletin* **76**, 253–263. doi:10.1016/j.brainresbull.2008.02.006

Kaplan G (2008b) The Australian magpie (*Gymnorhina tibicen*): an alternative model for the study of songbird neurobiology. In *The Neuroscience of Birdsong*. (Eds P Zeigler and P Marler) pp. 153–170. Cambridge University Press, Cambridge, UK.

Kaplan G (2009) Animals and music: between cultural definitions and sensory evidence. *Sign Systems Studies* **37**(3/4), 423–453. doi:10.12697/SSS.2009.37.3-4.03

Kaplan G (2011) Pointing gesture in a bird – merely instrumental or a cognitively complex behavior? *Current Zoology* **57**(4), 453–467. doi:10.1093/czoolo/57.4.453

Kaplan G (2014) Animal communication. *Wiley Interdisciplinary Reviews: Cognitive Science* **5**(6), 661–677. doi:10.1002/wcs.1321

Kaplan G (2015) *Bird Minds: Cognition and Behaviour of Australian Native Birds*. CSIRO Publishing, Melbourne.

Kaplan G (2016) Bird-brained and brilliant: Australia's avians are smarter than you think. *The Conversation*, 1 March, <https://theconversation.com/bird-brained-and-brilliant-australias-avians-are-smarter-than-you-think-51475>

Kaplan G (2017a) Audition and hemispheric specialization in songbirds and new evidence from Australian magpies. *Symmetry* **9**, 99. doi:10.3390/sym9070099

Kaplan G (2017b) Magpies can form friendships with people – here's how. *The Conversation*, 3 October, <http://theconversation.com/magpies-can-form-friendships-with-people-heres-how-83950#>.

Kaplan G (2018a) Babbling in a bird shows same stages as in human infants. The importance of the 'Social' in vocal development. *Trends in Developmental Biology* **10**, 97–123.

Kaplan G (2018b) Passerine cognition. In *Encyclopedia of Animal Cognition and Behavior*. (Eds J Vonk and TK Shackelford). Springer International Publishing, Switzerland (online). doi:10.1007/978-3-319-47829-6_875-1

Kaplan G (2018c) Development of meaningful vocal signals in a juvenile territorial songbird (*Gymnorhina tibicen*) and the dilemma of vocal taboos concerning neighbours and strangers. *Animal* **8**, 228. doi:10.3390/ani8120228

Kaplan G, Rogers LJ (2001) *Birds: Their Habit and Skills*. Allen & Unwin, Sydney.

Kaplan G, Rogers LJ (2004) Charles Darwin and animal behavior. In *Encyclopedia of Animal Behavior. 3 Vols*. (Eds M Bekoff and J Goodall) pp. 471–479, vol. 2 (introductory essay to vol. 2). Greenwood Publishing, Westport, CT.

Kaplan G, Rogers LJ (2013) Stability of referential signalling across time and locations: Testing alarm calls of Australian magpies (*Gymnorhina tibicen*) in urban and rural Australia and in Fiji. *PeerJ* **1**, e112. doi:10.7717/peerj.112

Kaplan G, Johnson G, Koboroff A, Rogers LJ (2009) Alarm calls of the Australian magpie (*Gymnorhina tibicen*): I. Predators elicit complex vocal responses and mobbing behaviour. *The Open Ornithology Journal* **2**, 7–16. doi:10.2174/1874453200902010007

Kearns AM, Joseph L, Cook LG (2013) A multilocus coalescent analysis of the speciational history of the Australo-Papuan butcherbirds and their allies. *Molecular Phylogenetics and Evolution* **66**, 941–952. doi:10.1016/j.ympev.2012.11.020

Keast A (1981) The evolutionary biogeography of Australian birds. In *Ecological Biogeography of Australia*. (Ed. A Keast) pp. 1584–1635. Dr W. Junk Publishers, The Hague, Netherlands.

Keast A, Recher HF, Ford H, Saunders DA, (Eds) (1985) *Birds of Eucalypt Forests and Woodlands: Ecology, Conservation and Management*. Surrey Beatty & Sons, Sydney.

Keedwell R, Sanders MD (1999) Australian magpie preys on banded dotterel chicks. *Notornis* **46**, 499–501.

Kelley LA, Healy SD (2011) Vocal mimicry. *Current Biology* **21**(1), R9–R10. doi:10.1016/j.cub.2010.11.026

Keywood MD, Hibberd MF, Emmerson KM (2017) *Australia State of the Environment 2016: Atmosphere*. Independent report to the Australian Government Minister for the Environment and Energy. Commonwealth of Australia.

King AS, McLelland J (1984) *Birds: Their Structure and Function*. Bailliere Tindall, London.

Kingma SA, Santema P, Taborsky M, Komdeur J (2014) Group augmentation and the evolution of cooperation. *Trends in Ecology & Evolution* **29**, 476–484. doi:10.1016/j.tree.2014.05.013

Klatt DH, Stefanski RA (1974) How does a mynah bird imitate human speech? *The Journal of the Acoustical Society of America* **55**(4), 822–832. doi:10.1121/1.1914607

Kleindorfer S, Hoi H, Fessl B (1996) Alarm calls and chick reactions in the moustached warbler, *Acrocephalus melanopogon*. *Animal Behaviour* **51**, 1199–1206. doi:10.1006/anbe.1996.0125

Koboroff A, Kaplan G (2006) Is learning involved in predator recognition? A preliminary study of the Australian magpie *Gymnorhina tibicen*. *Australian Field Ornithology* **23**, 36–41.

Koboroff A, Kaplan G, Rogers LJ (2008) Hemispheric specialization in Australian magpies (*Gymnorhina tibicen*) shown as eye preferences during response to a predator. *Brain Research Bulletin* **76**, 304–306. doi:10.1016/j.brainresbull.2008.02.015

Koboroff A, Kaplan G, Rogers LJ (2013) Clever strategists: Australian Magpies vary mobbing strategies, not intensity, relative to different species of predator. *PeerJ* **1**, e56. doi:10.7717/peerj.56

Komdeur J (1992) Importance of habitat saturation and territory quality for evolution of cooperative breeding in the Seychelles warbler. *Nature* **358**, 493–495. doi:10.1038/358493a0

Konishi M (1965) The role of auditory feedback in the control of vocalization in the white-crowned sparrow. *Zeitschrift für Tierpsychologie* **22**, 770–783.

Konishi M (1985) From behavior to neuron. *Annual Review of Neuroscience* **8**, 125–170. doi:10.1146/annurev.ne.08.030185.001013

Konishi M, Akutagawa E (1985) Neuronal growth, atrophy and death in a sexually dimorphic song nucleus in the zebra finch brain. *Nature* **315**, 145–147. doi:10.1038/315145a0

Kreisfeld R (1997) *Injuries involving magpies.* Research Centre for Injury Studies, Flinders University, Adelaide: <https://www.aihw.gov.au/reports/injury/injuries-involving-magpies/contents/table-of-contents>

Kroodsma DE (1996) Ecology of passerine song development. In *Ecology and Evolution of Acoustic Communication in Birds.* (Eds DE Kroodsma and EH Miller) pp. 3–19. Comstock Publishing (Cornell University Press), Ithaca, NY.

Kroodsma DE, Byers BE (1991) The function of birdsong. *American Zoologist* **31**, 318–328. doi:10.1093/icb/31.2.318

Kroodsma DE, Miller E, Ouellet H (Eds) (1982) *Acoustic Communication in Birds: Song Learning and Its Consequences.* Academic Press, New York.

Kroodsma DE, Vielliard JME, Stiles FG (1996) Study of bird sounds in the neotropics: urgency and opportunity. In *Ecology and Evolution of Acoustic Communication in Birds.* (Eds DE Kroodsma and EH Miller) pp. 269–281. Comstock Publishing (Cornell University Press), Ithaca, NY.

Kumar A (2003) Acoustic communication in birds. *Resonance* **8**(6), 44–55.

La Sorte FA, Lepczyk C, Aronson MF, Goddard MA, Hedblom M, Katti M, MacGregor-Fors I, Mörtberg U, Nilon CH, Warren PS, Williams NS (2018) The phylogenetic and functional diversity of regional breeding bird assemblages is reduced and constricted through urbanization. *Diversity and Distributions* **24**, 928–938. doi:10.1111/ddi.12738

Langmore NE, Feeney WE, Crowe-Riddell H, Luan H, Louwrens KM, Cockburn A (2012) Learned recognition of brood parasitic cuckoos in the superb fairy-wren *Malurus cyaneus. Behavioral Ecology* **23**(4), 798–805. doi:10.1093/beheco/ars033

Larkins D (1980) Selective food gathering by Australian magpie. *Corella* **4**(2), 36.

Latham J (1802) *Supplementum indicis ornithologici, sive systematis ornithologiae.* Prostat apud G. Leigh, J. et S. Sotheby, London.

Leach JA (1914a) The myology of the Bell-Magpie (Strepera) and its position in classification. *Emu* **14**(1), 2–38.

Leach JA (1914b) The R.A.O.U. 'Check-list'. *Emu* **13**(3), 190.

Leboucher G, Kreutzer ML, Dittami J (1994) Copulation solicitation displays in female canaries (*Serinus canaria*): are estradiol implants necessary? *Ethology* **97**, 190–197.

Lefebvre L, Gaxiola A, Dawson S, Timmermans S, Rosza L, Kabai P (1998) Feeding innovations and forebrain size in Australian birds. *Behaviour* **135**, 1077–1097. doi:10.1163/156853998792913492

Lefebvre L, Ducatez S, Audet JN (2016) Feeding innovations in a nested phylogeny of Neotropical passerines. *Philosophical Transactions of the Royal Society B* **371**(1690), 20150188.

Legge S (2004) *Kookaburra: king of the bush.* CSIRO Publishing, Melbourne.

Leonard H (1978) Forest Raven caching food. *Australian Bird Watcher* **7**, 212.

Lewis CF (1978) Little Raven caching food. *Australian Bird Watcher* **7**, 272.

Ley AJ (1995) Recovery of cached food by Torresian crow. *Australian Bird Watcher* **16**, 77.

Liebal K, Call J (2012) The origins of non-human primates' manual gestures. *Philosophical Transactions of the Royal Society B* **367**(1585), 118–128. doi:10.1098/rstb.2011.0044

Linacre E (1999) The last Ice Age in Australia, New Zealand, and Papua New Guinea. <http://www-das.uwyo.edu/~geerts/cwx/notes/chap15/lgm_oz.html>.

Lines WJ (1991) *Taming the Great South Land: A History of the Conquest of Nature in Australia*. Allen & Unwin, Sydney.

Low T (2003) *The New Nature*. Penguin Australia, Camberwell, Vic.

Luck GW, Possingham HP, Paton DC (1999) Bird responses at inherent and induced edges in the Murray Mallee, South Australia. 1. Difference in abundance and diversity. *Emu* **99**, 157–169. doi:10.1071/MU99019

Lundberg S (1985) The importance of egg hatchability and nest predation in clutch size evolution in altricial birds. *Oikos* **45**, 110–117. doi:10.2307/3565228

MacDougall-Shackleton SA, Ball GF (1999) Comparative studies of sex differences in the song-control system of songbirds. *Trends in Neurosciences* **22**, 432–436. doi:10.1016/S0166-2236(99)01434-4

MacLean EL, Hare B, Nunn CL, Addessi E, Amici F, Anderson RC, Aureli F, Baker JM, Bania AE, Barnard AM, Boogert NJ (2014) The evolution of self-control. *Proceedings of the National Academy of Sciences of the United States of America* **111**, 2140–2148. doi:10.1073/pnas.1323533111

Magat M, Brown C (2009) Laterality enhances cognition in Australian parrots. *Proceedings of the Royal Society of London B* **276**(1676), 4155–4162.

Magnotti JF, Katz JS, Wright AA, Kelly DM (2015) Superior abstract-concept learning by Clark's nutcrackers (*Nucifraga columbiana*). *Biology Letters* **11**(5), 20150148. doi:10.1098/rsbl.2015.0148

Manegold A (2008) Composition and phylogenetic affinities of vangas (Vangidae, Oscines, Passeriformes) based on morphological characters. *Journal of Zoological Systematics and Evolutionary Research* **46**, 267–277. doi:10.1111/j.1439-0469.2008.00458.x

Margoliash D, Fortune ES (1992) Temporal and harmonic combination-sensitive neurons in the zebra finch's HVc. *The Journal of Neuroscience* **12**, 4309–4326. doi:10.1523/JNEUROSCI.12-11-04309.1992

Margoliash D, Fortune ES, Sutter ML, Yu AC, Wren-Hardin BD, Dave AS (1994) Distributed representation in the song system of oscines: evolutionary implications and functional consequences. *Brain, Behavior and Evolution* **44**, 247–255. doi:10.1159/000113580

Marler P (1955) Characteristics of some animal calls. *Nature* **176**, 6–8. doi:10.1038/176006a0

Marler P (1991) Song-learning behavior: the interface with neuroethology. *Trends in Neurosciences* **14**, 199–206. doi:10.1016/0166-2236(91)90106-5

Marler P, Evans C (1996) Bird calls: just emotional displays or something more? *Ibis* **138**, 26–33. doi:10.1111/j.1474-919X.1996.tb04310.x

Marler P, Tamura M (1964) Culturally transmitted patterns of vocal behaviour in sparrows. *Science* **146**, 1483–1486. doi:10.1126/science.146.3650.1483

Martin AJ, Vickers-Rich P, Rich TH, Hall M (2014) Oldest known avian footprints from Australia: Eumeralla Formation (Albian) Dinosaur Cove, Victoria. *Palaeontology* **57**(1), 7–19. doi:10.1111/pala.12082

Martin TE (1992) Interaction of nest predation and food limitation in reproductive strategies. *Current Ornithology* **9**, 163–197. doi:10.1007/978-1-4757-9921-7_5

Marzluff JM, Walls J, Cornell HN, Withey J, Craig DP (2010) Lasting recognition of threatening people by wild American crows. *Animal Behaviour* **79**, 699–707. doi:10.1016/j.anbehav.2009.12.022

Mateos-Gonzalez F, Senar JC (2012) *Melanin-based trait predicts individual exploratory behaviour in siskins, Carduelis spinus. Animal Behaviour* **83**, 229–232.

Mathews GM (1908) Handlist of the birds of Australasia: With an introductory letter. *Emu* **7**, 1–108 [Supplement].

Mathews GM (1912) Reference list to the birds of Australia. *Novitates Zoologicae* **18**, 171–446.

McCaskill LW (1945) Preliminary report on the present position of the Australian magpies (*Gymnorhina tibicen*) in New Zealand. *New Zealand Bird Notes* **1**, 68–104.

McCasland JS, Konishi M (1981) Interaction between auditory and motor activities in an avian song control nucleus. *Proceedings of the National Academy of Sciences of the United States of America* **78**, 7815–7819. doi:10.1073/pnas.78.12.7815

McGraw KJ, Nogare MC (2005) Distribution of unique red feather pigments in parrots. *Biology Letters* **1**(1), 38–43. doi:10.1098/rsbl.2004.0269

McKechnie AE, Wolf BO (2010) Climate change increases the likelihood of catastrophic avian mortality events during extreme heat waves. *Biology Letters* **6**, 253–256. doi:10.1098/rsbl.2009.0702

McKechnie AE, Hockey PAR, Blair O, Wolf BO (2012) Feeling the heat: Australian landbirds and climate change. *Emu* **112**, i–vii. doi:10.1071/MUv112n2_ED

McKenzie R, Andrew RJ, Jones RB (1998) Lateralization in chicks and hens: new evidence for control of response by the right eye system. *Neuropsychologia* **36**(1), 51–58. doi:10.1016/S0028-3932(97)00108-5

McKinney F, Derrickson SR, Mineau P (1983) Forced copulation in waterfowl. *Behaviour* **86**, 250–293. doi:10.1163/156853983X00390

McNamara JM, Forslund P (1996) Divorce rates in birds: predictions from an optimisation model. *American Naturalist* **147**, 609–640. doi:10.1086/285869

Meade J, Nam KB, Beckerman AP, Hatchwell BJ (2010) Consequences of "load-lightening" for future indirect fitness gains by helpers in a cooperatively breeding bird. *Journal of Animal Ecology* **79**, 529–537. doi:10.1111/j.1365-2656.2009.01656.x

Mees GF (1964) Notes on two small collections of birds from New Guinea. *Zoologische Verhandelingen* **66**, 1–37.

Mees GF (1982) Birds from the lowlands of southern New Guinea (Merauke and Koembe). *Zoologische Verhandelingen* **19**, Monograph (188 pp).

Meltzoff AN, Brooks R (2007) Eyes wide shut: the importance of eyes in infant gaze following and understanding other minds. In *Gaze Following: Its Development and Significance*. (Eds R Flom, K Lee and D Muir) pp. 217–241. Erlbaum, Mahwah, NJ.

Meredith P (2017) Feathered geniuses. *Australian Geographic* Sep/Oct Issue, 96–103. <https://www.australiangeographic.com.au/topics/wildlife/2017/11/bird-intelligence/>

Milius S (1998) When birds divorce: who splits, who benefits, and who gets the nest. *Science News* **153**, 153–156. doi:10.2307/4010154

Milligan AW (1903) Description of a new Gymnorhina, with observations on *G. dorsalis*, Campbell: with plates. *Emu* **3**(2), 96–102. doi:10.1071/MU903096

Mirville MO (2013) The effect of group size on the cognitive abilities of the Australian magpie (*Cracticus tibicen dorsalis*). Honours thesis. University of Western Australia, Perth.

Mirville MO, Kelley JL, Ridley AR (2016) Group size and associative learning in the Australian magpie (*Cracticus tibicen dorsalis*). *Behavioral Ecology and Sociobiology* **70**, 417–427. doi:10.1007/s00265-016-2062-x

Mo M, Waterhouse DR, Hayler P, Hayler A (2016) Observations of mobbing and other agonistic responses to the Powerful Owl *Ninox strenua*. *Australian Zoologist* **38**(1), 43–51. doi:10.7882/AZ.2015.033

Moeed A (1976) Birds and their food resources at Christchurch International Airport, New Zealand. *New Zealand Journal of Zoology* **3**, 373–390. doi:10.1080/03014223.1976.9517926

Mohr BAR, Wähnert V, Lazarus D, Frey FA, Coffin MF, Wallace PJ, *et al.* (Eds) (2002) Mid-Cretaceous paleobotany and palynology of the central Kerguelen Plateau, southern Indian Ocean (ODP Leg 183, Site 1138). *Proceedings of the Ocean Drilling Program, Scientific Results 183*. Ocean Drilling Program Publications.

Mooney R (2009) Neurobiology of song learning. *Current Opinion in Neurobiology* **19**, 654–660. doi:10.1016/j.conb.2009.10.004

Mooney R (2014) Auditory–vocal mirroring in songbirds. *Philosophical Transactions of the Royal Society B* **369**, 20130179. doi:10.1098/rstb.2013.0179

Moore BR (1996) The evolution of imitative learning. In *Social Learning in Animals: The Roots of Culture*. (Eds CM Heyes and BG Galef, Jr) pp. 245–265. Academic Press, San Diego.

Morand-Ferron J, Lefebre L, Reader SM, Sol D, Elvin S (2004) Dunking behaviour in Carib grackles. *Animal Behaviour* **68**(6), 1267–1274.

Morand-Ferron J, Sol D, Lefebvre L (2007) Food stealing in birds: brain or brawn? *Animal Behaviour* **74**, 1725–1734. doi:10.1016/j.anbehav.2007.04.031

Morand-Ferron J, Quinn JL (2011) Larger groups of passerines are more efficient problem solvers in the wild. *Proceedings of the National Academy of Sciences of the United States of America* **108**(38), 15898–15903.

Morand-Ferron J, Cole EF, Rawles JEC, Quinn JL (2011) Who are the innovators? A field experiment with two passerine species. *Behavioral Ecology* **22**, 1241–1248. doi:10.1093/beheco/arr120

Morand-Ferron J, Cole EF, Quinn JL (2016) Studying the evolutionary ecology of cognition in the wild: a review of practical and conceptual challenges. *Biological Reviews of the Cambridge Philosophical Society* **91**, 367. doi:10.1111/brv.12174

Morris D (1956) The feather postures of birds and the problem of the origin of social signals. *Behaviour* **9**, 75–111. doi:10.1163/156853956X00264

Mortimer N, Campbell HJ, Tulloch AJ, King PR, Stagpoole VM, Wood RA, Rattenbury MS, Sutherland R, Adams CJ, Collot J, Seton M (2017) Zealandia: Earth's hidden continent. *Geological Society of America* **27**(3), 27–35.

Moyle RG, Cracraft J, Lakim M, Nais J, Sheldon FH (2006) Reconsideration of the phylogenetic relationships of the enigmatic Bornean Bristlehead (*Pityriasis gymnocephala*). *Molecular Phylogenetics and Evolution* **39**, 893–898. doi:10.1016/j.ympev.2006.01.024

Müller RD, Gaina C, Clark S (2000) Seafloor spreading around Australia. In *Billion-year Earth History of Australia and Neighbours in Gondwanaland*. (Ed. JJ Veevers) pp. 18–28. Gemoc Press, North Ryde, NSW.

Muñoz AP, Kéry M, Martins PV, Ferraz G (2018) Age effects on survival of Amazon forest birds and the latitudinal gradient in bird survival. *The Auk: Ornithological Advances* **135**, 299–313. doi:10.1642/AUK-17-91.1

Mya Mya Nu (1974) Vocal and auditory communication in parrots: some anatomical and behavioural aspects. PhD thesis, Zoology, University of New England, Armidale, NSW.

Mykytowycz R, Davies DW (1959) *Pasteurella pseudotuberculosis* in the Australian black-backed magpie *Gymnorhina tibicen* (Latham). *CSIRO Wildlife Research* **4**(1), 61–68. doi:10.1071/CWR9590061

Nef S, Allaman I, Fiumelli H, De Castro E, Nef P (1996) Olfaction in birds – differential embryonic expression of nine putative odorant receptor genes in the avian olfactory system. *Mechanisms of Development* **55**, 65–77. doi:10.1016/0925-4773(95)00491-2

Nelson DA (1999) Ecological influences on vocal development in the white-crowned sparrow. *Animal Behaviour* **58**, 21–36. doi:10.1006/anbe.1999.1118

Newton I (Ed.) (1989) *Lifetime Reproduction in Birds*. Academic Press, London.

Nguyen JMT, Worthy TH, Boles WE, Hand SJ, Archer M (2013) A new cracticid (Passeriformes: Cracticidae) from the Early Miocene of Australia. *Emu* **113**, 374–382. doi:10.1071/MU13017

Ngwenya A, Nahirney J, Brinkman B, Williams L, Iwaniuk AN (2017) Comparison of estimates of neuronal number obtained using the isotropic fractionator method and unbiased stereology in day old chicks (*Gallus domesticus*). *Journal of Neuroscience Methods* **287**, 39–46. doi:10.1016/j.jneumeth.2017.05.025

Norman JA, Ericson PG, Jønsson KA, Fjeldså J, Christidis L (2009) A multi-gene phylogeny reveals novel relationships for aberrant genera of Australo-Papuan core Corvoidea and polyphyly of the Pachycephalidae and Psophodidae (Aves: Passeriformes). *Molecular Phylogenetics and Evolution* **52**, 488–497. doi:10.1016/ j.ympev.2009.03.019

Noske RA (1985) Habitat use by three bark-foragers of eucalypt forests. In *Birds of Eucalypt Forests and Woodlands: Ecology, Conservation, Management*. (Eds A Keast, HF Recher, H Ford and D Saunders) pp. 193–204. RAOU and Surrey Beatty & Sons, Sydney.

Nottebohm F (1970) Ontogeny of bird song. *Science* **167**, 950–956.

Nottebohm F (1971) Neural lateralization of vocal control in a passerine bird. I. Song. *The Journal of Experimental Zoology* **177**, 229–261. doi:10.1002/jez.1401770210

Nottebohm F (1972) Neural lateralization of vocal control in a passerine bird. II. Subsong, calls and a theory of vocal learning. *The Journal of Experimental Zoology* **179**, 35–50. doi:10.1002/jez.1401790104

Nottebohm F (1975) Vocal behaviour in birds. In *Avian Biology, Vol. V.* (Eds DS Farner, JR King and KC Parkes) pp. 289–332. Academic Press, New York.

Nottebohm F (1977) Asymmetries in neural control of vocalization in the canary. In *Lateralization in the Nervous System*. (Ed. S Harnard) pp. 23–44. Academic Press, New York.

Nottebohm F (1980) Brain pathways for vocal learning in birds: a review of the first ten years. *Progress in Psychobiology and Physiological Psychology* **9**, 85–125.

Nottebohm F, Alvarez-Buylla A, Cynx J, Kirn J, Ling C-Y, Nottebohm M, Sutter R, Tolles A, Williams H (1990) Song learning in birds: the relation between perception and production. *Proceedings of the National Academy of Sciences of the United States of America* **329**, 115–124.

Nowicki S (1987) Vocal tract resonance in oscine bird sound production: evidence from birdsongs in a helium atmosphere. *Nature* **325**, 53–55. doi:10.1038/325053a0

Nowicki S, Capranica RR (1986) Bilateral syringeal interaction in the vocal production of an oscine bird. *Science* **231**, 1297–1299. doi:10.1126/science.3945824

Nowicki S, Peters S, Searcy WA, Clayton C (1999) The development of within-song type variation in song sparrows. *Animal Behaviour* **57**(6), 1257–1264. doi:10.1006/anbe.1999.1098

O'Connor RJ (1984) *The Growth and Development of Birds*. John Wiley & Sons, Chichester, England.

O'Leary RA, Jones DN (2002) Foraging by suburban Australian magpies during dry conditions. *Corella* **26**(2), 53–54.

O'Loghlen AL, Beecher MD (1999) Mate, neighbour and stranger songs: a female song sparrow perspective. *Animal Behaviour* **58**(1), 13–20. doi:10.1006/anbe.1999.1125

O'Neill MG, Taylor RJ (1984) Co-operative hunting by pied currawongs *Strepera graculina*. *Corella* **8**, 95–96.

Ocklenburg S, Ströckens F, Güntürkün O (2013) Lateralisation of conspecific vocalisation in non-human vertebrates. *Laterality* **18**, 1–31. doi:10.1080/1357650X.2011.626561

Ocklenburg S, Hirnstein M, Beste C, Güntürkün O (2014) Lateralization and cognitive systems. *Frontiers in Psychology* **5**, 1143. doi:10.3389/fpsyg.2014.01143

Odom KJ, Hall ML, Riebel K, Omland KE, Langmore NE (2013) Female song is widespread and ancestral in songbirds. *Nature Communication* **5**, 3379. doi:10.1038/ncomms4379

Olkowicz S, Kocourek M, Lucan RK, Porteš M, Fitch WT, Herculano-Houzel S, Némec P (2016) Birds have primate-like numbers of neurons in the forebrain. *Proceedings of the National Academy of Science* **113** (26), 7255–7260. doi:10.1073/pnas.1517131113

Olsen P (2001) *Feather and Brush: Three Centuries of Australian Bird Art*. CSIRO Publishing, Melbourne.

Panchanathan K, Frankenhuis WE (2016) The evolution of sensitive periods in a model of incremental development. *Proceedings of the Royal Society B* **283**(1823), 20152439.

Park TJ, Dooling RJ (1985) Perception of species-specific contact calls by budgerigars (*Melopsittacus undulatus*). *Journal of Comparative Psychology* **99**, 391–402. doi:10.1037/0735-7036.99.4.391

Parrish GR, Lock JW (1997) Classified summarised notes. North Island 1 July 1993 to 30 June 1996. *Notornis* **42**, 145–173.

Parsons FE (1968) Pterylography. The feather tracts of Australian birds with notes and observations, *Occasional Papers in Zoology No. 1*. Library Board of South Australia, Adelaide.

Paton DC (1977) Magpies attacking blackbirds. *South Australian Ornithologist* **27**, 185.

Pellis SM (1981a) A description of social play by the Australian magpie *Gymnorhina tibicen* based on Eshkol-Wachman notation. *Bird Behavior* **3**(3), 61–79. doi:10.3727/015613881791560685

Pellis SM (1981b) Exploration and play in the behavioural development of the Australian magpie *Gymnorhina tibicen*. *Bird Behavior* **3**(1–2), 37–49. doi:10.3727/015613881791560900

Pellis SM (1983) Development of head and foot coordination in the Australian magpie *Gymnorhina tibicen*, and the function of play. *Bird Behaviour* **4**(2), 57–62.

Pellis S, Pellis C (2009) *The Playful Brain. Venturing to the Limits of Neuroscience*. One World Publications, Oxford, UK.

Pellis SM, Burghardt GM, Palagi E, Mangel M (2015) Modeling play: distinguishing between origins and current functions. *Adaptive Behavior* **23**(6), 331–339.

Pepperberg IM (1987) Interspecies communication: a tool for assessing conceptual abilities in the African Grey parrot (*Psittacus erithacus*). In *Language, Cognition, Consciousness: Integrative Levels*. (Eds G Greenberg and E Tobach) pp. 31–56. Erlbaum, Hillsdale, NJ.

Pepperberg IM (2009) *The Alex Studies: Cognitive and Communicative Abilities of Grey Parrots*. Harvard University Press, Cambridge, MA.

Pettigrew JD, Konishi M (1976) Neurones selective for orientation and binocular disparity in the visual Wulst of the barn owl (*Tyto alba*). *Science* **193**, 675–678. doi:10.1126/science.948741

Phelan L (2018) Terror in the trees: Sydney's angriest magpie revealed. *Sun-Herald* 2 Oct. <https://www.smh.com.au/national/terror-in-the-trees-sydney-s-angriest-magpie-revealed-20180822-p4zyyx.html>

Pika S, Bugnyar T (2011) The use of referential gestures in ravens (*Corvus corax*) in the wild. *Nature Communications* **2**, 560. doi:10.1038/ncomms1567

Pike KN (2016) How much do helpers help? Variation in helper behaviour in the cooperatively breeding Western Australian magpie. PhD thesis, The University of Western Australia, Perth.

Pohl-Apel G, Sossinka R, Wyndham E (1982) Gonadal cycles of wild budgerigars, *Melopsittacus undulatus* (Psittaciormes: Platycercidae). *Australian Journal of Zoology* **30**(5), 791–797. doi:10.1071/ZO9820791

Poiani A (1993) Small clutch sizes as a possible adaptation against ectoparasitism: a comparative analysis. *Oikos* **57**, 237–240.

Poiani A, Pagel M (1997) Evolution of avian cooperative breeding: comparative tests of the nest predation hypothesis. *Evolution* **51**, 226–240. doi:10.1111/j.1558-5646.1997.tb02404.x

Porter D (1993) Magpie kills and eats a skylark. *Notornis* **40**, 246.

Potti J (1999) Maternal effects and the pervasive impact of nestling history on egg size in a passerine bird. *Evolution* **53**, 279–285. doi:10.1111/j.1558-5646.1999.tb05353.x

Potti J, Merino S (1996) Causes of hatching failure in pied flycatchers. *The Condor* **98**, 328–336. doi:10.2307/1369151

Prather JF, Peters S, Nowicki S, Mooney R (2008) Precise auditory–vocal mirroring in neurons for learned vocal communication. *Nature* **451**, 305–310. doi:10.1038/nature06492

Prior H, Güntürkün O (2001) Parallel working memory for spatial location and food-related object cues in foraging pigeons: binocular and lateralized monocular performance. *Learning & Memory* **8**, 44–51. doi:10.1101/lm.36201

Pullar EM (1932) Pseudotuberculosis of sheep due to *B pseudotuberculosis rodentium* (so called 'pyaemic hepatitis'). *Australian Veterinary Journal* **8**, 181–183. doi:10.1111/j.1751-0813.1932.tb00980.x

Qadri MA, Leonard K, Cook RG, Kelly DM (2018) Examination of long-term visual memorization capacity in the Clark's nutcracker (*Nucifraga columbiana*). *Psychonomic Bulletin & Review* (Feb), 1–7.

QNPWS (Queensland National Parks and Wildlife Service) (1993) *Living with Wildlife: The Magpie*. Department of Environment and Heritage, Queensland, Brisbane.

Randler C, Braun M, Lintker S (2011) Foot preferences in wild-living ring-necked parakeets (*Psittacula krameri*, Psittacidae). *Laterality* **16**, 201–206. doi:10.1080/13576500903513188

Reader's Digest (2002) *Reader's Digest Complete Book of Australian Birds*. Reader's Digest, Sydney.

Recher HF (1993) *Ground-dwelling and Ground-foraging Birds: The Next Round of Extinction?* University of New England, Armidale, NSW. <155.187.2.69/biodiversity/publications/articles/biolink4.html>.

Recher HF, Lim L (1990) A review of current ideas of the extinction, conservation and management of Australia's terrestrial vertebrate fauna. *Proceedings of the Ecological Society of Australia* **16**, 287–301.

Reiner A, Perkel DJ, Mello CV, Jarvis ED (2004) Songbirds and the revised avian brain nomenclature. In *Behavioural Neurobiology of Birdsong. Annales Vol. 1016.* (Eds HP Zeigler and P Marler) pp. 77–108. New York Academy of Sciences, New York.

Ricklefs RE (1993) Sibling competition, hatching asynchrony, incubation period, and lifespan in altricial birds. In *Current Ornithology*. pp. 199–276. Springer, Boston, MA.

Ricklefs RE, Starck JM (1998) Embryonic growth and development. In *Avian Growth and Development*. (Eds RE Ricklefs and JM Starck) pp. 31–58. Oxford University Press, New York.

Ricklefs RE, Austin SH, Robinson WD (2017) The adaptive significance of variation in avian incubation periods. *The Auk* **134**(3), 542–550.

Ridley J, Douglas WY, Sutherland WJ (2005) Why long-lived species are more likely to be social: the role of local dominance. *Behavioral Ecology* **16**, 358–363. doi:10.1093/beheco/arh170

Riebel K (2009) Song and female mate choice in zebra finches: A review. In *Advances in the Study of Behavior*. pp. 197–238. Elsevier Academic Press, San Diego.

Riehl C, Strong MJ (2015) Social living without kin discrimination: experimental evidence from a communally breeding bird. *Behavioral Ecology and Sociobiology* **69**, 1293–1299. doi:10.1007/s00265-015-1942-9

Rivera-Cáceres KD, Templeton CN (2017) A duetting perspective on avian song learning. *Behavioural Processes*, online early. doi:10.1016/j.beproc.2017.12.007

Robin L (2001) *The Flight of the Emu: A Hundred Years of Australian Ornithology 1901–2001*. Melbourne University Press, Melbourne.

Robinson A (1956) The annual reproductive cycle of the magpie, *Gymnorhina dorsalis* Campbell, in south-western Australia. *Emu* **56**, 233–336. doi:10.1071/MU956233

Robinson FN (1975) Vocal mimicry and the evolution of birdsong. *Emu* **75**, 23–27. doi:10.1071/MU9750023

Robinson FN, Curtis HS (1996) The vocal displays of the lyrebirds (Menuridae). *Emu* **96**, 258–275. doi:10.1071/MU9960258

Rogers LJ (1980) Lateralisation in the avian brain. *Bird Behavior* **2**, 1–12. doi:10.3727/015613880791573835

Rogers LJ (1995) *The Development of Brain and Behaviour in the Chicken*. CAB International, Oxford, UK.

Rogers LJ (1997) *Minds of Their Own: Thinking and Awareness in Animals*. Allen & Unwin, Sydney.

Rogers LJ (2007) Development of functional lateralization in the avian brain. *Brain Research Bulletin* **76**, 304–306.

Rogers LJ, Andrew R (Eds) (2002) *Comparative Vertebrate Lateralization*. Cambridge University Press, Cambridge, UK.

Rogers LJ, Anson JM (1979) Lateralisation of function in the chicken forebrain. *Pharmacology, Biochemistry, and Behavior* **10**, 679–686. doi:10.1016/0091-3057(79)90320-4

Rogers LJ, Bradshaw JL (1996) Motor asymmetries in birds and nonprimate mammals. In *Manual Asymmetries in Motor Performance*. (Eds D Elliott and EA Roy) pp. 3–31. CRC Press, New York.

Rogers LJ, Kaplan G (2000) *Songs, Roars and Rituals: Communication in Birds, Mammals and Other Animals*. Harvard University Press, Boston.

Rogers LJ, Kaplan G (2003) *Spirit of the Wild Dog*. Allen & Unwin, Sydney.

Rogers LJ, Kaplan G (Eds) (2004) *Comparative Vertebrate Cognition: Are Primates Superior to Non-primates*. Kluwer Academic, New York.

Rogers LJ, Kaplan G (2006) An eye for a predator: lateralization in birds, with particular reference to the Australian magpie (*Gymnorhina tibicen*). In *Behavioral and Morphological Asymmetries in Vertebrates*. (Eds Y Malashichev and W Deckel) pp. 47–57. Landes Bioscience, Georgetown, TX.

Rogers LJ, Zucca P, Vallortigara G (2004) Advantage of having a lateralized brain. *Proceedings. Biological Sciences* **271**, S420–S422. doi:10.1098/rsbl.2004.0200

Rogers LJ, Vallortigara G, Andrew RJ (2013) *Divided Brains: The Biology and Behaviour of Brain Asymmetries*. Cambridge University Press, Cambridge, UK.

Rollinson DJ (2002) Food caching behaviour in the Australian Magpie (*Gymnorhina tibicen*). *The Sunbird* **32**(1), 19–21.

Rowley I (1972) *Bird Life*. Collins, Sydney.

Rowley I (1978) Communal activities among white-winged choughs *Corcorax melanorhamphus*. *Ibis* **120**(2), 178–197. doi:10.1111/j.1474-919X.1978.tb06774.x

Rowley ICR, Russell EM (2009) Family Artamidae (woodswallows). In *Handbook of the Birds of the World. Vol 14: Bush-shrikes to Old World Sparrows*. (Eds J del Hoyo, A Elliott and D Christie) pp. 286–307. Lynx Edicions, Barcelona.

RSPB (2009) *Kleptomaniacs of the skies*. 3 July. <https://goo.gl/V93Mp6>.

Rule H (2013) *Lesson 7: Dreaming Stories and Science – Why the magpie sings at dawn*. (Ed. Davidson) NSW Department of Education and Training. <https://online.det.nsw.edu.au/blog/226294-informationliteracyandnarrativestructurelessonsforyear7in2013/entry/lesson_7_dreaming_stories_and>.

Russell EM, Rowley ICR (2009) Family Cracticidae (butcherbirds). In *Handbook of the Birds of the World. Vol. 14: Bush-shrikes to Old World Sparrows*. (Eds J del Hoyo, A Elliott and D Christie) pp. 308–343. Lynx Edicions, Barcelona.

Sanderson K, Crouch H (1993) Vocal repertoire of the Australian Magpie *Gymnorhina tibicen* in South Australia. *Australian Bird Watcher* **15**(4), 162–164.

Sasaki A, Sotnikova TD, Gainetdinov RR, Jarvis ED (2006) Social context-dependent singing-regulated dopamine. *The Journal of Neuroscience* **26**, 9010–9014. doi:10.1523/JNEUROSCI.1335-06.2006

Schmidt LG, Bradshaw SD, Follett BK (1991) Plasma levels of luteinizing hormone and androgens in relation to age and breeding status among cooperatively breeding Australian magpies (*Gymnorhina tibicen* Latham). *General and Comparative Endocrinology* **83**, 48–55. doi:10.1016/0016-6480(91)90104-E

Schmidt MF, Wild JM (2014) The respiratory-vocal system of songbirds: Anatomy, physiology, and neural control. *Progress in Brain Research* **212**, 297–335. doi:10.1016/B978-0-444-63488-7.00015-X

Schneider DM, Mooney R (2015) Motor-related signals in the auditory system for listening and learning. *Current Opinion in Neurobiology* **33**, 78–84. doi:10.1016/j.conb.2015.03.004

Schodde R, Mason IJ (1999) *The Directory of Australian Birds: Passerines*. CSIRO Publishing, Melbourne.

Schodde R, Dickinson EC, Steinheimer FD, Bock WJ (2010) The date of Latham's Supplementum Indicis Ornithologici: 1801 or 1802? *South Australian Ornithologist* **35**(8), 231–235.

Schuppli C, Graber SM, Isler K, van Schaik CP (2016) Life history, cognition and the evolution of complex foraging niches. *Journal of Human Evolution* **92**, 91–100.

Schwarz MP, Fuller S, Tierney SM, Cooper SJB (2006) Molecular phylogenetics of the exoneurine allodapine bees reveal an ancient and puzzling dispersal from Africa to Australia. *Systematic Biology* **55**(1), 31–45. doi:10.1080/10635150500431148

Secomb D (2005) Arboreal foraging and food-caching by the Forest Raven *Corvus tasmanicus*. *Australian Field Ornithology* **22**, 67–71.

Sharpe RB (2008) *A Hand-list of the Genera and Species of Birds (Vol. 3)*. BiblioBazaar, Charleston, SC.

Sherry DF (2017) Food storing and memory. In *Avian Cognition*. (Eds C ten Cate and SD Healy) pp. 52–74. Cambridge University Press, Cambridge, UK.

Sherry DF, Schacter DL (1987) The evolution of multiple memory systems. *Psychological Review* **94**(4), 439–454. doi:10.1037/0033-295X.94.4.439

Shurcliffe A, Shurcliffe K (1974) Territory in the Australian magpie (*Gymnorhina tibicen*): an analysis of its size and change. *South Australian Ornithologist* **26**, 127–132.

Sibley CG, Ahlquist JE (1985) The phylogeny and classification of the Australo-Papuan passerine birds. *Emu* **85**, 1–14. doi:10.1071/MU9850001

Similä T, Ugarte F (1993) Surface and underwater observations of cooperatively feeding killer whales in northern Norway. *Canadian Journal of Zoology* **71**(8), 1494–1499. doi:10.1139/z93-210

Simpson HB, Vicario DS (1990) Brain pathways for learned and unlearned vocalizations differ in zebra finches. *The Journal of Neuroscience* **10**, 1541–1556. doi:10.1523/JNEUROSCI.10-05-01541.1990

Simpson K, Day N, Trusler P (1993) *Field Guide to the Birds of Australia*. 4th edn. Penguin, Giraween, New South Wales.

Sinervo B (1997–2006) *Optimal Foraging Theory: Constraints and Cognitive Processes*. University of Southern California, Santa Cruz: available at <http://bio.research.ucsc.edu/~barrylab/classes/CHAPTER_PDFS/Chap_6_Optimal.pdf>, pp. 105–130.

Slabbekoorn H (2017) Animal communication: competition for acoustic space in birds and fish. In *Biocommunication: Sign-Mediated Interactions between Cells and Organisms*. (Eds R Gordon and J Seckbach) pp. 327–363. World Scientific, Singapore.

Slater PJB (1989) Bird song learning: causes and consequences. *Ethology Ecology and Evolution* **1**, 19–46. doi:10.1080/08927014.1989.9525529

Smith CC, Reichman OJ (1984) The evolution of food caching by birds and mammals. *Annual Review of Ecology and Systematics* **15**, 329–351. doi:10.1146/annurev.es.15.110184.001553

Soha JA, Peters S (2015) Vocal learning in songbirds and humans: a retrospective in honor of Peter Marler. *Ethology* **121**, 933–945. doi:10.1111/eth.12415

Stacey PB, Koenig WD (1990) *Cooperative Breeding in Birds: Long-Term Studies of Ecology and Behavior*. Cambridge University Press, Cambridge, UK.

Stamps JA, Krishnan VV (1999) A learning-based model of territory establishment. *The Quarterly Review of Biology* **74**, 291–318. doi:10.1086/393163

Stefanski RA, Falls JB (1972) A study of distress calls of song, swamp and whitethroated sparrows (Aves: Fringillidae). I. Intraspecific responses and functions. *Canadian Journal of Zoology* **50**, 1501–1512. doi:10.1139/z72-199

Sternberg H (1989) Pied Flycatcher. In *Lifetime Reproduction in Birds*. (Ed. I Newton) pp. 55–74. Academic Press, London.

Stevens JR, King AJ (2013) The lives of others: social rationality in animals. In *Simple Heuristics in a Social World*. (Eds R Hertwig, U Hoffrage and the ABC Research Group) pp. 409–431. Oxford University Press, Oxford, UK.

Sulikowski D, Burke D (2011) Movement and memory: different cognitive strategies are used to search for resources with different natural distributions. *Behavioral Ecology and Sociobiology* **65**, 621–631. doi:10.1007/s00265-010-1063-4

Suthers RA (2001) Peripheral vocal mechanism in birds: are songbirds special? *Netherlands Journal of Zoology* **51**(2), 217–242. doi:10.1163/156854201X00288

Suthers RA, Goller F, Hartley RS (1994) Motor dynamics of song production by mimic thrushes. *Journal of Neurobiology* **25**(8), 917–936. doi:10.1002/neu.480250803

Suthers RA, Goller F, Hartley RS (1996) Motor stereotypy and diversity in songs of mimic thrushes. *Journal of Neurobiology* **30**, 231–245. doi:10.1002/(SICI)1097-4695(199606)30:2<231::AID-NEU5>3.0.CO;2-6

Suthers R, Wild M, Kaplan G (2011) Mechanisms of song production in the Australian magpie. *Journal of Comparative Physiology. A, Neuroethology, Sensory, Neural, and Behavioral Physiology* **197**(1), 45–59. doi:10.1007/s00359-010-0585-6

Sutter E (1951) *Growth and Differentiation of the Brain in Nidifugous and Nidicolous Birds.* Almqvist & Wiksell, Uppsala.

Sutton AO, Strickland D, Norris DR (2016) Food storage in a changing world: implications of climate change for food-caching species. *Climate Change Responses* **3**(1), 12. doi:10.1186/s40665-016-0025-0

Suzuki TN (2016) Semantic communication in birds: evidence from field research over the past two decades. *Ecological Research* **31**, 307–319. doi:10.1007/s11284-016-1339-x

Tarwater CE, Ricklefs RE, Maddox JD, Brawn JD (2011) Pre-reproductive survival in a tropical bird and its implications for avian life histories. *Ecology* **92**(6), 1271–1281. doi:10.1890/10-1386.1

Taylor H (2011) Anecdote and anthropomorphism: writing the Australian pied butcherbird. *Australasian Journal of Ecocriticism and Cultural Ecology* **1**(1), 20.

Taylor H (2017) *Is Birdsong Music? Outback Encounters with an Australian Songbird.* Indiana University Press, Bloomington, IN.

Tchernichovski O, Marcus G (2014) Vocal learning beyond imitation: mechanisms of adaptive vocal development in songbirds and human infants. *Current Opinion in Neurobiology* **28**, 42–47. doi:10.1016/j.conb.2014.06.002

Teather KL (1992) An experimental study of competition for food between male and female nestlings of the red-winged blackbird. *Behavioral Ecology and Sociobiology* **31**, 81–87. doi:10.1007/BF00166340

Temeles EJ (1994) The role of neighbours in territorial systems: when are they 'dear enemies'? *Animal Behaviour* **47**, 339–350. doi:10.1006/anbe.1994.1047

ten Cate C (2014) On the phonetic and syntactic processing abilities of birds: from songs to speech and artificial grammars. *Current Opinion in Neurobiology* **28**, 157–164. doi:10.1016/j.conb.2014.07.019

Thoburn L (1978) The vocal mimicry of the superb lyrebird. Unpublished honours thesis, University of Sydney, Sydney.

Thoma P, Bauser DS, Norra C, Brüne M, Juckel G, Suchan B (2014) Do you see what I feel? – Electrophysiological correlate of emotional face and body perception in schizophrenia. *Clinical Neurophysiology* **125**(6), 1152–1163.

Thomas L (2000) The human dimensions of an extreme human–wildlife interaction: Australian magpies in suburbia. PhD thesis, Griffith University.

Thorn S, Werner SA, Wohlfahrt J, Bässler C, Seibold S, Quillfeldt P, Müller J (2016) Response of bird assemblages to windstorm and salvage logging – insights from analyses of functional guild and indicator species. *Ecological Indicators* **65**, 142–148. doi:10.1016/j.ecolind.2015.06.033

Thorpe WH (1964) Mimicry, vocal. In *A New Dictionary of Birds.* (Ed. AL Thomson) Nelson, London.

Tommasi L, Vallortigara G (2004) Hemisphere processing of landmark and geometric information in male and female domestic chicks (*Gallus gallus*). *Behavioural Brain Research* **155**, 85–96. doi:10.1016/j.bbr.2004.04.004

Toon A, Mather PB, Baker AM, Durrant KL, Hughes JM (2007) Pleistocene refugia in an arid landscape: analysis of a widely distributed Australian passerine. *Molecular Ecology* **16**, 2525–2541. doi:10.1111/j.1365-294X.2007.03289.x

Tramontin AD, Hartman VN, Brenowitz EA (2000) Breeding conditions induce rapid and sequential growth in adult avian song control circuits. *The Journal of Neuroscience* **20**(2), 854–861. doi:10.1523/JNEUROSCI.20-02-00854.2000

Treisman M (1978) Bird song dialects, repertoire size, and kin association. *Animal Behaviour* **26**, 814–817. doi:10.1016/0003-3472(78)90146-X

Tremont S (1995) Australian magpie chases and kills southern whiteface. *Australian Birdwatcher* **16**(2), 79.

Umbers KD, Mappes J (2015) Post-attack deimatic display in the mountain katydid, Acripeza reticulata. *Animal Behaviour* **100**, 68–73. doi:10.1016/j.anbehav.2014.11.009

Vallentin D, Kosche G, Lipkind D, Long MA (2016) Inhibition protects acquired song segments during vocal learning in zebra finches. *Science* **351**(6270), 267–271. doi:10.1126/science.aad3023

Vallortigara G (1992) Right hemisphere advantage for social recognition in the chick. *Neuropsychologia* **30**(9), 761–768. doi:10.1016/0028-3932(92)90080-6

Vallortigara G, Rogers LG, Bisazza A (1999) Possible evolutionary origins of cognitive brain lateralization. *Brain Research. Brain Research Reviews* **30**, 164–175. doi:10.1016/S0165-0173(99)00012-0

Vallortigara G, Cozzutti C, Tommasi L, Rogers LJ (2001) How birds use their eyes: opposite left-right specialisation for the lateral and frontal visual hemifield in the domestic chick. *Current Biology* **11**(1), 29–33. doi:10.1016/S0960-9822(00)00027-0

Vallortigara G, Snyder A, Kaplan G, Bateson P, Clayton NS, Rogers LJ (2008) Are animals autistic savants? *PLoS Biology* **6**(2), e42. doi:10.1371/ journal.pbio.0060042

van Asten T, Hall ML, Mulder RA (2016) Who cares? Effect of coping style and social context on brood care and defense in superb fairy-wrens. *Behavioral Ecology* **27**, 1745–1755. doi:10.1093/beheco/arw096

van den Brink V, Dolivo V, Falourd X, Dreiss AN, Roulin A (2012) Melanic color dependent antipredator behavior strategies in barn owl nestlings. *Behavioral Ecology* **23**(3), 473–480. doi:10.1093/beheco/arr213

Vander Wall SB, Balda RP (1977) Coadaptations of Clark's nutcracker and the pinyon pine for efficient seed harvest and dispersal. *Ecological Monographs* **47**, 89–111. doi:10.2307/1942225

Veltman CJ (1984) The social system and reproduction in a New Zealand magpie population, and a test of the cooperative breeding hypothesis. PhD thesis, Department of Zoology, Massey University, Palmerston North, NZ.

Veltman CJ (1989) Flock, pair and group-living lifestyles without cooperative breeding by Australian magpies Gymnorhina tibicen. *Ibis* **131**, 601–608. doi:10.1111/j.1474-919X.1989.tb04795.x

Veltman CJ, Carrick R (1990) Male-biased dispersal in Australian magpies. *Animal Behaviour* **40**, 190–192. doi:10.1016/S0003-3472(05)80682-7

Veltman CJ, Hickson RE (1989) Predation by Australian magpies (*Gymnorhina tibicen*) on pasture invertebrates: are non-territorial birds less successful? *Australian Journal of Ecology* **14**(3), 319–326. doi:10.1111/j.1442-9993.1989.tb01440.x

Venz S (2016) *Pretty with a purpose: the anatomy of a feather.* <http://www.wideopenpets.com/pretty-with-a-purpose-the-anatomy-of-a-feather>.

Vestjens WJM, Carrick R (1974) Food of the black-backed magpie, *Gymnorhina t. tibicen*, at Canberra. *Australian Wildlife Research* **1**, 71–83. doi:10.1071/WR9740071

Vicario DS (1994) Motor mechanisms relevant to auditory-vocal interactions in songbirds. *Brain, Behavior and Evolution* **44**, 265–278. doi:10.1159/000113581

Vicario DS, Yohay KH (1993) Song-selective auditory input to a forebrain vocal control nucleus in the zebra finch. *Journal of Neurobiology* **24**, 488–505. doi:10.1002/neu.480240407

Vickers-Rich P, Baird RF, Monaghan J, Rich TH (Eds) (1991) *Vertebrate Palaeontology of Australasia*. Thomas Nelson, Melbourne.

Vince MA (1973) Some environmental effects on the activity and the development of the avian embryo. In *Behavioural Embryology*. (Ed. G. Gottlieb) pp. 285–323. Academic Press, New York.

Visser GH (1998) Development of temperature regulation. In *Avian Growth and Development*. (Eds RE Ricklefs and JM Starck) pp. 117–156. Oxford University Press, New York.

Waite ER (1903) Sympathetic song in birds. *Nature* (68), 322.

Ward M, Schlossberg S (2004) Conspecific attraction and the conservation of territorial songbirds. *Conservation Biology* **18**(2), 519–525. doi:10.1111/j.1523-1739.2004.00494.x

Warne RM, Jones DN, Astheimer LB (2010) Attacks on humans by Australian magpies (*Cracticus tibicen*): territoriality, brood-defence or testosterone? *Emu* **110**(4), 332–338. doi:10.1071/MU10027

Weary D, Krebs J (1987) Birds learn song from aggressive tutors. *Nature* **329**, 485.

Weathers WW, Schoenbaechler DC (1976) Regulation of temperature in the Budgerygah, *Melopsittacus undulatus*. *Australian Journal of Zoology* **24**, 39–47. doi:10.1071/ZO9760039

West MJ, King AP (1990) Mozart's Starling. *American Scientist* **78**, 106–113.

Whitfield MC, Smit B, McKechnie AE, Wolf BO (2015) Avian thermo-regulation in the heat: scaling of heat tolerance and evaporative cooling capacity in three southern African arid-zone passerines. *The Journal of Experimental Biology* **218**(11), 1705–1714. doi:10. 1242/jeb.121749

Wikelski M, Spinney L, Schelsky W, Scheuerlein A, Gwinner E (2003) Slow pace of life in tropical sedentary birds: a common-garden experiment on four stonechat populations from different latitudes. *Proceedings. Biological Sciences* **270**(1531), 2383–2388. doi:10.1098/rspb.2003.2500

Wild JM (1994) Visual and somatosensory inputs to the avian song system via nucleus uvaeformis (Uva) and a comparison with the projections of a similar thalamic nucleus in a nonsongbird, *Columba livia*. *The Journal of Comparative Neurology* **349**, 512–535. doi:10.1002/cne.903490403

Wild JM (1997) Neural pathways for the control of birdsong production. *Journal of Neurobiology* **33**, 653–670. doi:10.1002/(SICI)1097-4695(19971105)33:5<653::AID-NEU11>3.0.CO;2-A

Wild JM, Kubke MF, Peña JL (2008) A pathway for predation in the brain of the barn owl (*Tyto alba*). *The Journal of Comparative Neurology* **509**(2), 156–166. doi:10.1002/cne.21731

Wiley D, Ware C, Bocconcelli A, Cholewiak D, Friedlaender A, Thompson M, Weinrich M (2011) Underwater components of humpback whale bubble-net feeding behaviour. *Behaviour* **148**, 575–602.

Williams JM, Slater PJB (1990) Modelling bird song dialects: the influence of repertoire size and numbers of neighbours. *Journal of Theoretical Biology* **145**, 487–496. doi:10.1016/S0022-5193(05)80483-7

Williams TD (1994) Intraspecific variation in egg size and egg composition in birds: effects on offspring fitness. *Biological Reviews of the Cambridge Philosophical Society* **68**, 35–59. doi:10.1111/j.1469-185X.1994.tb01485.x

Wise KK, Conover MR, Knowlton FF (1999) Response of coyotes to avian distress calls: testing the startle-predator and predator-attraction hypotheses. *Behaviour* **136**, 935–949. doi:10.1163/156853999501658

Wittenberger JF, Hunt GL, Jr (1985) The adaptive significance of coloniality in birds. *Avian Biology* **8**, 1–78.

Wood SR, Sanderson KJ, Evans CS (2000) Perception of terrestrial and aerial alarm calls by honeyeaters and falcons. *Australian Journal of Zoology* **48**, 127–134. doi:10.1071/ZO99020

Woxvold IA, Mulder RA, Magrath MJL (2006) Contributions to care vary with age, sex, breeding status and group size in the cooperatively breeding apostlebird. *Animal Behaviour* **72**, 63–73. doi:10.1016/j.anbehav.2005.08.016

Wrangham RW (1981) Drinking competition in vervet monkeys. *Animal Behaviour* **29**(3), 904–910. doi:10.1016/S0003-3472(81)80027-9

Wright J, Berg E, de Kort SR, Khazin V (2001) Safe selfish sentinels in a cooperative bird. *Journal of Animal Ecology* **70**, 1070–1079. doi:10.1046/j.0021-8790.2001.00565.x

Xie S, Turrell EJ, McWhorter TJ (2017) Behavioural responses to heat in captive native Australian birds. *Emu* **117**(1), 51–67. doi:10.1080/01584197.2016

Zahavi A, Zahavi A (1997) *The Handicap Principle*. Oxford University Press, Oxford.

Zinkivskay A, Nazir F, Smulders TV (2009) What–Where–When memory in magpies (*Pica pica*). *Animal Cognition* **12**(1), 119–125. doi:10.1007/s10071-008-0176-x

Index

Aboriginal Dreaming, *see* Australian Aboriginal and Torres Strait Islander peoples
Aboriginal names for magpies   2
Acanthorhynchus tenuirostris (eastern spinebill)   81
Accipiter fasciatus (goshawk)   52, 82, 93, 125, 157, 195
Acripeza reticulata (mountain katydid) (diet source)   67, 69
acrobatics
    adult play behaviour   146
    aerial display   89
Acrocephalus palustris (migratory marsh warbler), vocal mimicry   172
Acrocephalus sechellensis (Seychelle warbler)   85
Aegithinidae (ioras)   5
aerial bluff, as defence strategy   89
affiliation closeness, communication and   180
affiliative cues   184
Africa   4
    Australian magpies and   19
    land bridge to Australia   8, 9, 11, 12
    songbirds   5–6
African bees, Australia and   6–7
African bush-shrikes (Malaconotidae)   6, 8, 11
African grey parrot (*Psittacus erithacus*) mimicry capacity   173, 176
African Plate   6
age limits, song learning   169
ages of the Earth   4–14
    see also Cretaceous; Jurassic; K-T boundary; Miocene; Tertiary
agonistic behaviour patterns   182, 183, 204
airflow (phonation)   160–1, 163
air sacs   159
airspace rules   91–2
airstream (vocalisation)   160
alarm calls   157, 181, 191–2
    learnt   169
    referential   192–7
    social rules and   153
    songs and   180
allopreening   184
'all purpose' beaks   29
    see also beak
Alps   102
altitude, airspace corridors   91

altricial species, egg hatching   116, 117–18
Americas   112
amphibians (diet source)   59
amplitude modulation, magpie song   163–4
anatomy   27–39
ancestry, magpies and lyrebirds   174
anger expression   184
anisodactyl (unequal toes) feet   38
ankle (intertarsal joint)   37
Antarctica   4, 7
Antarctic Plate   6, 8
Anthochaera carunculata (red wattlebird)   95, 111
anti-predator behaviour   82
    nestlings and   123
    pointing posture   196–7
ants (diet source)   59
apostlebird (*Struthidea cinerea*), cooperation behaviour   72, 156
Aquila audax (wedge-tailed eagle)   51, 52, 82, 93, 157
architecture, magpie nests   106
Area X   165, 166
Armidale   25, 53, 87, 93, 191
Artamidae (woodswallows)   4, 5, 6, 17, 29, 34
    brain mass/body weight ratio   45
Artamus leucorynchus (white-breasted woodswallow)   34
Ask a Biologist   30
asymmetry, syringeal muscles   161
asynchronous hatching, predatory species   118
attrition rate, offspring   113–14
audition   52–4
auditory perception
    lateralisation and   52–5
    scarab larvae hunting and   53–4
Australia   3, 4
    land bridges and   6, 7–8, 9
    magpie distribution   20
    New Guinea links   8–14
Australian Aboriginal and Torres Strait Islander peoples
    magpies and   1, 10
Australian magpie (*Gymnorhina tibicen*)   5, 11
    Aboriginal Dreaming and   1–2
    brain mass/body weight ratio   44–5
    European settlement and   213
    humans and   199–212
    lyrebird mimicry comparison   174–6

translocation (government policy) 210–11
vocal mimicry examples 173
see also subspecies, Australian magpie
Australian Plate 6
Australo-Papuan
    bellbirds (Oreoicidae) 5
    butcherbirds 11
    centred fantails (Rhipiduridae) 6
    corvida 174
avian tree of life 4

babbling, humans 173
backyard feeding, caching and 67
banding records, dispersal 131–2
barbed wire 125, 200
barking owl 110
Bassian origin, magpies 23
batiss (Platysteiridae) 5
beak 29
    as tool 128
    as weapon 29
    colour 29
    song action and 217
    types 29
beak-clapping 156, 185, 209
beak length measurement, nestling development 124
bears, *see* extractive foraging skills
bees
    colonisation from Africa 6–7, 8, 9
    diet source for magpies 68–9
    hemispheric lateralisation 47
beetles (diet source) 59
begging calls 181
    fledglings 124
    hatchlings 119–20
    nestlings 168, 169
behavioural variations, magpie subspecies 23–4
Bergman's rule 22
bilateral mini-breath 162
bill length variations, magpie subspecies 23
binder twine injuries 105–8
binocular vision 48, 49–50, 128
biological parenthood, social parenthood and 102
bipedalism, rarity in animals 37–8
bird eggshells (diet source) 59
BirdLife Australia v, 15, 17, 115, 132, 133
bird of paradise (Paradiseidae) 5, 6
'Bird of Prey' 14, 15
birds of prey 129
    defence against 156–7
    extractive skills 60
    feet 39
    nests 105

referential calls and 193–6
black-backed magpie (BBM) 17–18, 20, 21, 23
    defence 87, 195
    lack of sexual dimorphism 101
    nestling weight increase 124
black-billed magpie (*Pica hudsonia*) 12
black butcherbird (*Melloria quoyi*) 10, 11, 12, 13
    brain mass/body weight ratio 45
    colouration 30–1
    intercontinental travel 10
    magpies and 17
black-capped vireo (*Vireo atricapilla*) 86
black kite (*Milvus migrans*) 72–3, 147
black-shouldered kite (*Elanus axillaris*), as predator 93
'blind areas' 48
boatbills (Machaerirynchidae) 5
body mass, brain weight and 44–5
body posture
    communication and 182–3
    singing and 141
body size
    brain size and 43–4
    debate about 22
body temperature, nestlings and 122
body weight
    egg size and 116
    neuron density and 47
bonding
    breeding behaviours and 66, 99–114
    duetting 189
bone marrow extraction 61–2
Bornean bristlehead (*Pityriasis gymnocephala*) 5
boundary conflict 88
bowerbirds (Ptilonorhynchidae)
    brain mass/body weight ratio 46
    mobbing behaviour and 95
    play behaviour 147
brain architecture 42, 52
brain development 41–55
    compared to mammals 155
    fledglings 126–7
branch forks, nests in 105
branchlings 125–6, 217
breeding females, friction 150
breeding pairs, percentage of magpie population 99
breeding season 24
    bonding and 99–114
    colouration and 31
    disconnection from vocalisations 159
    extended 102–3
    preparation for 203–5
    public interest 115

swooping behaviour   199
territory defence   80, 90
Brisbane   21, 24
Brisbane City Council, magpie territory
   warning   208
broadband vocalisations   162
Broken Ridge   7, 8, 9
bronchi   162
brooding, clutch size and   110–11
brooding female, territory defence and   90
Broome   3
brown goshawk (*Accipiter fasciatus*), *see*
   goshawk
brown thrasher (*Toxostoma rufum*), two-voice
   vocalisation   163
browsing behaviour   221
Brunette Downs   24
bubble-netting   75
budgerigar (*Melopsittacus undulatus*)
   brain lateralisation   50
   cooperative behaviour   155
   non-hierarchical feeding   119
   opportunistic breeding   103
   thermoregulation   76
   vocal mimicry   173
   water sources   91
bush-shrikes (Malaconotoidea)   4
butcherbirds (Cracticidae)   4, 5, 11, 17, 34
   beak   29
   brain mass   44, 45
   caching   58, 67
   feeding canopies   63
   hopping   37
   relatedness to magpies   11
   rescue of nestlings   207–8
   skill levels   62
   threat from   208

caching, hoarding and   58, 64–7
   see also food retrieval, storage
call to arms   181, 183
Calyptorhynchus latirostris (Carnaby's black
   cockatoo)   76
camouflage
   feathers and   32
   food foraging and   59
Campephagidae (cuckooshrikes, trillers)   5
canary (*Serinus canaria domestica*), song
   control system   164, 165, 166
Canberra   24, 25, 85, 86, 87, 101, 111, 113, 132,
   154, 173, 191
Cape York Peninsula   10, 19
capuchin monkeys, food calls   74
Carnaby's black cockatoo (*Calyptorhynchus
   latirostris*)   76
carolling   186–8

defence strategy   87, 88, 89
duetting and   189–90
juvenile listening to   171
musculature and   164, 165
social rules   153
carpometacarpal bones and phalanges
   (manus)   32
carrion crow (*Corvus corone corone*), group
   structures   134–5
carrion (diet source)   59
cassowary, *see* southern cassowary
cats, nest building and   127
Central Australia   77, 100
central Panama   113
centre-edge effect (territory claims)   84
cerebellum   42
cervical vertebrae   36
cetaceans
   systematic foraging   75
   vocal mimicry   172
Chabert vanga (*Leptopterus chabert*)   34
channel-billed cuckoo (*Scythrops
   novaeholladiae*), as parasite   111
chickens   39
   brain lateralisation   48, 50
children, aggression to magpies   206
chimpanzees
   brain size   155
   extractive skills   60
   pointing behaviour   197
chorus carolling   188
   see also carolling; song; songbirds
Clark's nutcracker (*Nucifraga
   columbiana*)   64, 65
cleanliness   106, 141
climate change, non-perishable foods and   67
climate factors
   breeding season and   102, 113
   feeding patterns and   75–8
closed beak singing   141–3
clutch frequency, latitude and   112
clutch size, brooding and   110–11
cockatoos
   brain mass/body weight ratio   46
   extractive skills   60
   social play   148
cockerels, referential calls   74, 192–3
cockroaches (diet source)   60
Coffs Harbour   25, 87, 101, 191
cognitive capacity   154–5
   brain lateralisation and   54
   brain mass and   46
   communication   94, 180
   corvids   47
   foraging   57–78
   mimicry   176–7

pointing behaviour   197
colonisation, bees from Africa   6–7, 8, 9
colour-associated foraging tasks   179
colouration, breeding period   31
common (or eastern) koel (*Eudynamys orientalis*)   111
communication systems   179–98
compatibility factors   112
complex skill niches   61
complex song   168
compulsory bicycle helmets, swoopings and   209
computer systems, magpie brains and   41
concentrations, magpie dispersal   133
continental drying   23
continuity (singing)   141–3
contralateral hemispheric lateralisation, vision   48, 49, 50
control system, magpie song   164–7
convergence, duetting   190
convergent evolution, brain capacity   42
Coober Pedy   3
Coolgardie, WA, *see* Aboriginal names for magpies
Coolup   25
cooperative behaviour   72–5, 155–7
    breeding   96, 101, 102, 156
coordinated flying skills   95
copulation   104
Corcorax melanorhamphos (white-winged chough)   5, 61, 119
Corvidae (ravens, crows and others)   4, 5, 6
    agonistic behaviour   183
    brain mass/body weight ratio   45
    cognitive capacities   47
    cooperative behaviour   155
    food caching   58
    Wulst area   42
Corvus corone corone (carrion crow), group structures   134–5
Corvus tasmanicus (forest raven), caching   67
courtship patterns   100, 103–5
Cracticidae (butcherbirds, currawongs, magpies)   5, 6, 10, 17
Cracticus nigrogularis (pied butcherbird), duetting with magpies   189–90
creation stories, magpies and   1–2
crepuscular foraging   64
crested berrypecker (Paramythiidae)   5
crested pigeon
    shared territories   81, 82
    wing flapping   185
Cretaceous period   2, 7
crimson rosella (*Platycercus elegans*), non-hierarchical feeding   119

cross-species play   148
crouching behaviour   151–2
crows (Corvidae)   5
    colouration   30
    food caching   58
    individual human recognition   205
    see also ravens
cuckoo shrike (Campephagidae)   5
    mobbing behaviour and   95
currawongs (Cracticidae)   4, 5, 11, 34, 44, 45, 105
    as prey   93
    feeding canopies   63
    food retrieval   58, 59, 61
    hopping   37
    manus flutter   182
    parasite species and   111
curtain of sound (eagle alarm call)   193–4
cyclists, swooping and   199, 206, 208

daily life   139–58
'dangerous bird' category   203, 206
Darwin, Charles   151, 154
dawn chorus, songbirds   143
'dear enemy' hypothesis (carolling)   87, 186
death rates, dispersal and   132
deciduous trees, nest sites   108
defence strategies   115–16, 204, 209
    predators and   92–7, 220
    territory size and   86–90
deforestation, food sources and   76
deimatic display   67
Descartes, René   154
development data
    eggs   116–18
    fledglings   116, 126
    foraging   72
    hatching   118–19, 124–7
    predation risk   129
    referential signalling   192
    song production   159–77, 180
    vocalisation   108
dialects   186
dicotyledons (diet source)   59
Dicruridae (drongo)   4, 5
dietary requirements   57–78
    hatchlings   119
digits   37, 38
dinosaurs   3
direct binocular gaze   184
directed song   181
direct magpie–human attacks, rogue magpies   200
discrimination tasks, learning of   129
disease/parasites   202

dispersal patterns 25
  juveniles 131–5
display activity, territory defence and 87
distress calls 180, 181, 191–2
  nestlings 169
distribution, magpies within Australia 20
'divorced' magpies 101
DNA fingerprinting, extra-pair paternity 102
dogs
  agonistic behaviour 183
  crouching behaviour 151
dominant females, breeding and 102
dominant males, feeding rules and 153
dopamine, singing-regulated 144
drongo (Dicruridae) 4, 5
drought 76, 90
duetting 188–91
Dumetella carolinensis (grey catbird), two-voice vocalisation 163
dunking behaviour 67, 69
duration, vocalisations 141–3
Dutch explorers, New Guinea magpies and 18

eagle 129
  asynchronous hatching 118
  referential calls by magpies 94, 193–5, 196–7
Early Miocene 10
earthworms (diet source) 59, 60
  hatchling diet 119, 120
east coast, dispersal patterns 132
east coast–west coast magpie links 10
East Gondwana 3
eastern Australia 156
eastern Melbourne 85
eastern spinebills (*Acanthorhynchus tenuirostris*)
  shared territories 81
eastern yellow robin (*Eopsaltria australis*), mobbing behaviour 95, 204
eclectus parrot 10
ecological factors
  flight and 27–8
  song learning 171–2
  territory size and 86
ectoparasites 140
edible food items 70–1, 128
egg incubation, colouration and 32
egg shifting, cooperative behaviour 156
egg weight, hatchling survival rate and 116, 117
Elanus axillaris (black-shouldered kite) 93
Emblem status (Western Australia) 14
embryo development 117–18
emotional expression, vocalisations 180, 181

emus, brain size 44
endangered species, specialised feeders 213
environmental factors
  breeding success and 99
  melanin levels and 31–2
  song learning 170
Eocene 12
Eopsaltria australis (eastern yellow robin), mobbing behaviour 95, 204
episodic memory, food retrieval and 57
Eudynamys orientalis 111
Eulacestomidae (ploughbill) 5
eumelanic birds, *see* melanin
euphoria, singing and 144
Eurasia 3
Eurasian magpie (*Pica pica*) 12, 64
Europe 112
European corvids 46
European settlement
  dispersal patterns and 132
  landscape change and 75–6
  magpie success and 213
  naming of magpies 14
European starling (*Sturnus vulgaris*)
  song control system 165
  vocal mimicry 172
European stonechat (*Saxicola torquata*), survival rates 112
evaporative cooling, climate change and 76
evolution
  flight 27
  song behaviour 131
exploratory behaviour, melanin and 31
extractive foraging skills 60–2
  learning 128–9
extra-pair mating 102, 150
extreme heatwave events 76–8
eye/foot coordination, tool use and 61–2
eye-gaze following 197
eyelids 28

facial expressions, feather postures and 183–5
facial micro-signals, anger and 184
facial recognition 55, 205, 209–10
failure, nest building 106–7
Falco peregrinus (peregrine falcon), as predator 93, 129, 157
fantails (Rhipiduridae) 5
  brain mass/body weight ratio 46
  mobbing behaviour 95, 204
fearlessness, anti-predator behaviour 95, 96
feather postures, facial expressions and 183–5
feathers
  colour changes 30, 127
  fluffing 183–4

positioning 28
structure 30–4
wings and 33
feeding patterns 213
climate factors 75–8
non-hierarchical 119, 120–2
social rules 153
feet 38–9
female and male equality, magpie singing and 165–6
females
incubation period 117
courtship solicitation 32, 103
dispersal behaviour 132
nest building 105
nest control 116
nestlings and 122
territorial dominance 150
femur 36
feral animal attacks 202
fibula 36, 37
Ficedula hypoleuca (pied flycatcher) 101, 119
figbirds (Oriolidae) 5
figs (diet source) 59
Fiji 195–6, 222
magpie population 214
fine materials, nest building 105, 106
fine visual discrimination 55
flashpoints, swooping 209
fledgling
development 124–8
begging and 152
colouration 216
flight feathers (remiges) 32–3
flight
airspace rules 91
evolution of 27, 28
learning of 124–7, 203, 217
see also mobbing behaviour; swooping behaviour
flight paths, Africa to Australia 9
flocking birds 83, 134
flycatchers, territory establishment and 85
flying education, fledglings 124–6
food calls 74, 181
food learning behaviour 69–70, 126, 128–9
food nutritional value, magpie judgements 120–1
food retrieval 65
social rules 152
source identification 70–1
stealing 65
storage 62
food supplies, disruption 76
foot/brain coordination 39

foot digits, bone marrow extraction and 61–2
foraging behaviour
adaptability 213
colour-assisted 179
efficiency 69–72
heat stress and 77–8
hierarchy 63
niches 60
scarab larvae and 53–4
social behaviour and 72–5
forebrain (pallium) 42
song control system and 164–5
forest ravens (*Corvus tasmanicus*), caching 67
fossil record, Riversleigh 10
foxes, extractive skills 60
free-ranging birds, food caching 58–9
free-ranging magpies, visual capacity 50
friendships, magpies and humans 211–12
frogmouths, beak-clapping 185
frogs (diet source) 59
frugivorous birds, food retrieval 57–8
fuchsias, nectar 69
functionality, magpie brain 41
fundamental frequency, magpie song 162

galah
magpie alarm calls and 157
mimicry capacity 176
generic alarm call 191–2, 195
genetic 106, 124
assertiveness 102
help and feeding 69
relationships 102
signature 102
studies 17, 81
tests 156
gestures 94–5
gift-giving (courtship) 104–5
glossy black-cockatoo 213
Goldfields Aboriginal dialect 2
gonadal testosterone, nest defence and 210
Gondwana 3–4
goshawk (*Accipiter fasciatus*) 82
defence against 93, 157, 195
fledglings and 125
magpie perception of 52
government policy, magpie translocation 210–11
grains (diet source) 59
grass length, habitat quality and 25
grasshoppers (diet source) 59
gratification deferment, dunking and 67
greeting displays, returned fledglings 135
grey butcherbird 11
brain mass/body weight ratio 45

grey catbird (*Dumetella carolinensis*), two-voice vocalisation 163
grey-crowned babbler (*Pomatostomus temporalis*), cooperative behaviour 72, 156
Griffith University 102
ground feeding 32, 37, 63
   fledglings 126
group activities 144–5
group carolling, defence and 89, 186
group hierarchy, food caching and 65
group protection
   habitat quality and 153
   stability 72
group size
   communication and 180
   novel food exploration and 75
   territory size and 85
   variation 24–5
growth rate, hatchlings 120
Guardian, The v, 115
gular fluttering 76
Gulf of Mexico 2
gum trees, nest sites 108
Gymnorhina tibicen (Australian magpie) 11, 12, 15–16, 17, 25
   G. t. dorsalis (western magpie) 10, 18, 20, 22, 25, 81
   G. t. eylandtensis (Top End magpie) 18, 20–1, 22, 23, 116
   G. t. hypoleuca (Tasmanian magpie) 18, 20, 22
   G. t. longirostris (black-backed magpie) 10, 18, 20, 21, 22, 25
   G. t. papuana (New Guinea magpie) 10, 18, 20
   G. t. telonocua 20, 22, 25
   G. t. terraereginae 18, 20, 22
   G. t. tibicen 18, 20, 22, 81, 127, 135
   G. t. tyrannica 18, 20, 22, 23, 25, 116
   see also subspecies, Australian magpie

habitat quality
   destruction of 79
   group protection and 153
   group size variation and 25
hand-raised magpies, wild-raised magpies compared with 121
Harris's hawk (*Parabuteo unicinctus*) 72
Hartin, Dee and Tylah 209
hatching 115, 117, 118–19
hats, recognition of 209
Hawaii 6
hawks, as predators 129
head 34–5
health care 140–1

pair bonding and 101
Heard Island 7
hearing ability 181–2
heat control 28
heat stress 76, 77, 78
Haliaeetus leucogaster (white-bellied sea eagle) defence against 157, 220
helmetshrikes (Prionopidae) 4
helper magpies 123
helping-at-the-nest routine 72
hemispheric lateralisation, brain capacity 47–50, 51, 54
   see also lateralised functions; left-brain activity; right-hemisphere capacity
herons, dunking 67
Hieraaetus morphoides (little eagle), as predator 93
hierarchical feeding, hatchlings 119
hierarchy, friction and 150
high arousal calls 191–2
high vocal centre (HVC) 164–5, 166
hindbrain 42
hoarding, caching and 64
'holding court' 154, 155
Holocene Epoch 2
'homeless' pairs 66
homesickness, magpie carolling and 187
honey bees (diet source) 68, 69
honeyeaters
   brain mass/body weight ratio 46
   communication systems 179
   food retrieval 57
   mobbing behaviour and 96
hooded butcherbird 11
hopping 37
hormones, song and 165
hot dry weather, dispersal and 133
housekeeping (nests) 122–3
Hughes, Jane 25, 83, 102
human-based components, nest building 105
human brain, magpie brain and 43
human home ownership, magpie territories and 79
humans
   face recognition of 127
   hemispheric lateralisation 48
   infant memory capacity 176–7
   magpies and ix, 199–212
   territorial destruction and 82
   vocal mimicry 172–3
hummingbirds, food retrieval 57
hyperpallium (Wulst area) 42–3

inanimate objects, lyrebird mimicry 174
incubation time 117

India   4
Indian Ocean   6, 7, 8, 9
individual compatibility   112
individual human recognition   205
individual singers, carolling and   186
individual-specific pair preference   101
inedible food items   70–1
infertile eggs   117
in-group friction   150–5
injury rescue   211
innate vocalisations   169
innovation
   feeding skills   60–1
   song   180
   nest building   106
insects (diet source)   58, 59, 64
insulation, feathers   30
integument   28–9
intentional assessment, food hoarding and   65
intentional communications   181
inter-clavicular sac   160
inter-individual competition, food caching and   65
intertarsal joint (ankle)   37
intimidation displays   33, 149
intruders
   airspace rules and   91–2
   defence against   87
invertebrates (diet source)   59
invisible obstructions, vocalisation quality   182
ioras (Aegithinidae)   4, 5
irritation, facial expressions   185
Isotropic Fractionator   46

Javan Trough   6
jays, food hoarding   65
jewel-babblers (Psophodidae)   5
Jurassic period   2, 10
juveniles
   adult abandonment   201, 202–3
   affiliation   184
   colour variations   216
   duetting   188–90
   food learning behaviours   69–70
   human interaction with   205
   object play   219
   play behaviour   64, 66, 146–7, 211
   social punishments and   151–2
   vulnerability to attack   32, 129

Kazakhstan   112
Keast, Allan   63
keel bone   36
Kentucky Creek   90

Kenya   91, 112
Kerguelen Plateau   7, 8, 9
Kimberley   10, 11
kleptomaniac species   64
knee joint   37
kookaburra
   asynchronous hatching   118
   courtship   104
   foraging   63
   laughter   186
   mimicry capacity   169, 175
   mobbing behaviour   95
   play behaviour   147
   syrinx   160
   territory establishment   85
   window of return   136
koolbardie (Noongar name)   1–2
K-T boundary   2

labia   160, 162
lace monitor, magpie perception of   52
land bridges   6
land exposure, sea-level changes and   10
landing learning, fledglings   124–5
landscape change, European settlement and   75–6
language, referential alarm calls and   192–7
Laniidae (shrikes)   5
lapwing (*Vanellus miles*), mobbing behaviour and   95, 204
larder hoarders   64
larvae (diet source)   59, 60
   see also scarab larvae
larynx   159, 160
   singing and   142–3
late Eocene   6
lateralised functions, scarab larvae identification and   60
lateral magnocellular nuclei, anterior neostriatum (lMAN)   165, 166
latitudinal gradients, offspring survival rates   112
Launceston   25
leadership display, as defence strategy   89
learning process
   airspace rules   91
   extractive foraging   128–9
   flight   124–7, 203, 217
   bone marrow extraction   62
   vocalisations   181
left-brain activity   50, 155
left-ear specialisation, foraging and   54
left-eye visualisation, predators and   51
left-side syrinx   161–2, 163–4
legs   36, 37–8

leisure activity, magpie birdsong as 143–4
leisure months, daily life 140
Leptopterus chabert (Chabert vanga) 34
life span 112, 113
lifetime learning ability, song 170, 180
lions, teamwork 96
listening activity 53, 219
Little Desert National Park and nest positions 108
little eagle (*Hieraaetus morphoides*) 82
    defence against 52, 93, 157
little raven, caching 67
live prey (food source) 128
lizards 37, 59
local council officers, nestlings and 207–8
locust plague 64
loners, absence of 139
long-distance dispersal, juveniles 132–4
longevity, social intelligence and 179
long-necked waterbirds 36
lower leg 37
lumbar vertebrae 36
lumen 160
lyrebirds (Menuridae) 39
    vocal mimicry 173, 174–5

McDonald Island 7
Machaerirynchidae (boatbills) 5
Madagascar 4, 6, 8, 9, 34
magpie lark
    duetting calls 188
    mobbing behaviour and 96
Magpie Whisperer, The 38, 147
Malaconotidae (African bush-shrikes) 6
male marsh warbler, vocal mimicry 172
males
    brooding period and 115
    dispersal behaviour 132
    nest defence 206, 210
    non-brooding nest presence 110, 111, 117
Malurus cyaneus (superb fairy-wren), song function and 167–8
mammals
    body weight/neuronal density ratio 47
    hemispheric lateralisation 48
    magpie diet source 59
    play behaviour 146
    sound production 160
Manorina melanocephala (noisy miners)
    cooperation behaviour 72, 155
    magpie alarm calls and 157
    shared territories 81, 82
manus (carpometacarpal bones and phalanges) 32, 182
marbled frogmouth 10

marginal groups 83, 84
    airspace corridors and 91
masked lapwing (*Vanellus miles*)
    hatchling incubation 117
    mobbing behaviour and 96
mass extinction, Cretaceous period 2–3
mating display, feathers and 32
mating function, lyrebird vocalisations 175
mating procedure 104
maturation process 127
mealworms 69
meat scavenging 59
megapodes 39
melanin (pigment) 30–1
Melbourne 21, 24, 25, 87, 191
Melloria quoyi (black butcherbird) 11, 12, 13, 17
Melopsittacus undulatus (budgerigar), non-hierarchical feeding 119
membranes 160
memory ability 127, 159
Menuridae (lyrebirds) 39, 173, 174–5
metabolic costs, foraging activity 71
metabolism, hot climates and 112
mice (diet source) 59
microsite selection 76, 77
midbrain 42
mid-Cretaceous 8
midday singing peak 143
migratory birds 6
    magpie territoriality and 174
    non-migratory birds and 134
migratory marsh warbler (*Acrocephalus palustris*), vocal mimicry 172
millipedes (diet source) 60
Milvus migrans (black kite) 72–3
Milvus milvus (red kite) 64
mimicry 181
    song learning and 169, 172–6
Mimidae, two-voice vocalisation 163
mince meat 121
Miocene 12
mirror neurons, song learning and 171
misdemeanours 151
mixed referential calls 193
mobbing behaviour 199, 204
    anti-predator defence 95
    controlled use of 205
    learned 129
    warning to humans 206
mobile groups 83
mobility 134
Mohouidae (whiteheads) 5
Monarchidae (monarchs and allies) 4, 5, 6
monitor lizard (*Varanus varius*) 93, 109

monocular vision   49
monoculture, habitat destruction and   79
monogamous pairs   101
morphology and distribution, magpie subspecies   22
mother–daughter alliances   144–6
motor system (magpie brain)   55, 166
mottled whistler (Rhagologidae)   5
mountain katydid (*Acripeza reticulata*) (diet source)   67, 69
mountain peltops (*Peltops montanus*)   34
multimodal stimuli   53
multiple breeding events   81
multiple species shared territories   81–2
muscular system   28
    vocalisations   159, 164
music, carolling and   190–1

naming, Australian magpie   12
National Geographic   203
native trees, absence of   106
natural attrition   201, 202–3
natural habitat, brain lateralisation and   49
Natural History Museum (London)   14, 15
Nature   190
neck   35–6
nectar feeders   57, 213
negotiating display, territory defence   88
Neosittidae (sittellas)   5
neotropical passerine, survival rates   113
nervous system   41–55
nest building   105–8, 115, 156
nest cleanliness   122
nest defence
    predators and   108–10, 122–3
    vocalisations   181
nest position and predators   108–10
nesting trees, decline of   209
nestlings
    abandonment of   201, 202–3
    feather ruffling and   184
    food learning   120–1, 128
    group reintroduction   135–6
    infanticide of   210–11
    injuries – binder twine   107–8
    self-guided song learning   168, 171
    vocalisations   167
nest parasites   111
neuron density, measurement of   46, 47
neuroscience, sound production and   167
New Atlas   99
New Caledonian crows, stick use   58
New England Tableland   24, 73, 90, 93, 126, 135
    road kill statistics   201–2

New Guinea   6
    Australian links to   8–14
    Corvids and   12
New Guinea lowland peltops (*Peltops blainvillii*)   13
New Guinea magpie   18, 19–20
New South Wales   53, 64, 81, 93, 95, 101, 105, 108, 111, 124, 126, 143, 156, 173, 176
    road kill statistics   200–2
New Zealand   4, 6, 153, 214
New Zealand kea, brain mass/body weight ratio   45
niche balance, maintenance of   82
nictitating membrane   29
nidicolous organisms   126
nightingales, song learning   169
Ninox strenua (powerful owl), mobbing behaviour   95
nocturnal predators, nest sites and   110
nocturnal vocalisation   64
noisy friarbirds, mobbing behaviour and   95
noisy miners (*Manorina melanocephala*)
    cooperation behaviour   72, 155
    magpie alarm calls and   157
    shared territories   81, 82
nomadic species   134, 155
'non-aggressive' males   211
non-brooding males   111
non-directed song   181
non-hierarchical feeding, hatchlings   119, 120–2
non-migratory birds, migratory birds and   134
non-native trees, nest building failure   106–7
non-perishable foods, caching   67
non-reproducing individuals   99
non-vocal signals   182–4
North Africa   112
Northern Hemisphere, bird sound and   185
northern India   112
Northern Tableland   53, 126
Northern Territory   22, 24, 25, 116, 173, 221
north-western Australia, magpie migration   18–19
Nucifraga columbiana (Clark's nutcracker)   64, 65
nuclei (magpie brain), song production and   166
Nullarbor Plain   24, 25
numerical imbalance, territory defence and   87
nutrition
    diet and   62–3, 71
    song learning and   170

object play   147, 148, 219

offspring
    care of   96, 100, 115–37
    numbers (seasonal)   111
    scarab larvae hunting and   54
olfactory learning   53
open groups   83
opportunistic breeding   103
opposable digits   39
optic tectum   42
orang-utan
    dangers from   203–4
    see also extractive foraging skills   60
Oreoicidae (Australo-Papuan bellbirds)   5
Oriolidae (orioles, figbirds)   5
oscine species   130, 161
ostriches, cooperative behaviour   155
overall health, melanin levels and   31
overshooting, fledglings   125
overtones, carolling   190
overwintering, European stonechat   112
owls   129
    beak-clapping   185

Pachycephalidae (whistlers, shrike-thrushes and pitohuis)   5
Pacific region, magpie translocation to   214
Pacific (swamp) harrier, eagle alarm call and   195–6
paeleoanthropology, bird's legs and   37
pair bonding   100–2, 210
pallium (avian forebrain)   42
palm cockatoo   10, 44, 45
palm moths, attempted biological control   195
palm oil plantations   222
panting   76, 77
Papua-New Guinea/Australian links   8–14, 30
Papuan magpie   25
Parabuteo unicinctus (Harris's hawk)   72
Paradisaeidae (birds of paradise)   5, 6
parallel tones, pure tones and   160
parallel working memory, brain lateralisation and   54–5
Paramythiidae (tit berrypecker and crested berrypeckers)   5
pardalotes (Pardalotidae), shared territories   81
parental absence, song learning and   170–1
parental care, post hatching   119–22, 156
parrots (Psittaciformes)
    agonistic behaviour   183
    communication systems   179
    cooperative behaviour   155
    cuddling   144
    food manipulation   61
    social play   148

territory sharing   81, 82
    vocal mimicry   173
passerine birds (Passeridae)   4, 5, 6
    cooperative behaviour   155
    dispersal   132
    evolution of   10
    two-voice vocalisation   163
passers-by, mobbing behaviour and   205
pedestrian crowding, swooping and   209
peltops
    P. blainvillii (New Guinea lowland peltops)   11, 13
    P. montanus (mountain peltops)   34
Pepperberg, Irene   176
perching/songbirds (passeriformes)   5
peregrine falcon (*Falco peregrinus*), as predator   93
peripheral territories, survival rate   113
perishable foods   6, 65, 67
permanent territory groups   83–4, 86
personal conflict   151
Perth   25
pet cats   82
philopatric dispersal   132
phonation (airflow)   160–1, 162
physical cognition, nest building and   106
physical features, vocalisations   159
physical maturation, nestlings   123–4
Pica hudsonia (black-billed magpie)   12
Pica pica (Eurasian magpie)   12, 64
pied butcherbird (*Cracticus nigrogularis*)   11, 189–90
pied currawong
    brain mass/body weight ratio   45
    colouration   30
pied flycatcher (*Ficedula hypoleuca*)
    females   101
    non-hierarchical feeding   119
pigeons, hemispheric lateralisation   48, 54
pigment variations   19–20
Pilbara, fossil record   10
pine trees, nest sites   109
'Piping Roller'   14, 15
pitohuis (Pachycephalidae)   5
Pityriaseidae (Bornean bristlehead)   5
plant matter (diet source)   59
plasticity, song learning   169
Platycercus elegans (crimson rosella), non-hierarchical feeding   119
Platysteiridae (wattle-eyes and batiss)   5
playback, referential alarm call   194–5
play behaviours   66, 145–8, 211, 212
    cross-species play   148
    object play   219
    social play   147–8, 218

solitary play   146, 147, 148, 218
play fighting   148–50, 183
Pleistocene   11
ploughbill (Eulacestomidae)   5
plumage
  colour   87
  camouflage   110
  cleanliness   141
  variations   19–20, 21, 23
  white-backed and black-backed magpies   17
plural breeding season   103, 150
pointing behaviour, predator defence and   94, 196–7
  see also posture
poisoning, injuries to magpies   200
'police of the bush' (territorial defence)   157
Pomatostomus temporalis (grey-crowned babbler), cooperative nest-building   156
'Port Jackson Painter' (Watling)   14, 15
post-fledgling, vocal development   172
posture
  appeasement   182
  hanging heads   152
  pointing   196
  submissive   135–6, 151–2, 182
powerful owl (*Ninox strenua*), mobbing behaviour   95
pre-breeding season, birdsong and   141–2
precocial species, egg hatching   117
predator expulsion   89
predators
  asynchronous hatching   118
  defence techniques against   50–2, 87, 92–7
  distress calls and   191–2
  extractive skills   60
  learning behaviour and   129–30
  nest position and   108–10
preen (uropygial) gland   28, 35
prehistoric periods   2
pre-reproductive survival rates, western slaty antshrike   113
prickly pear (food source)   59
primary (central) system, vocalisations   159
primary feathers   33
primates
  agonistic behaviour   183
  brain capacity   47
  extractive skills   60
  gestures   94
  social play   148
princess parrots, water sources   91
problem solving   55, 62
profitability of prey (equation)   71
Psittacus erithacus, mimicry capacity   173, 176

Psophodidae (whipbirds, quail-thrushes and jewel-babblers)   5
Ptilonorhynchidae (bowerbirds)   46, 95, 147
punishment rules   151–5
pure tones, parallel tones and   160

quail-thrushes (Psophodidae)   5
Quaternary Ice Age   23
Queensland   3, 10, 18, 24, 25, 89, 173

rabbits, as prey   93
rain exposure, nests   109
raptors   39, 42, 43
rationality, animals and   154
ravens (Corvidae)   5
  as predators   108
  brain mass/body weight ratio   45–6
  colouration   30, 32
  communication systems   179
  dunking   67
  food hoarding   65
  gestures   94
  mobbing behaviour and   95, 96
  nests   105
  shuffling movement   37
  social play   148
recipients, vocalisations   182
rectrices (tail feathers)   33
red kite (*Milvus milvus*)   64
red wattlebird (*Anthochaera carunculata*)
  mobbing behaviour and   95
  parasite species and   111
referential vocalisations   181, 192–7
regional differences, defensive displays   89
reintroduction time limits, fledglings   135–6
relationship variability, territories and   80
reproductive ability
  reproductive behaviour versus   116
  song capacity and   42, 141
reptiles (diet source)   59
resource dependability, territory   85
respiratory process   159
Rhagologidae (mottled whistler)   5
rhinoceroses, dangers from   203
Rhipiduridae (fantails)   5, 6
ribs   37
right-hemisphere capacity   50, 52, 155
right-side syrinx   161–2, 163, 164
risk behaviour, vocalisations and   130
Riversleigh World Heritage Area   3, 10
road kill statistics   200–2
robins, extractive skills   60
robust nucleus of archistriatum (RA) (Area X)   165, 166
rogue magpies   200

rooks, body weight/neuronal density ratio   46, 47
rules suspensions, territory defence vigilance   90–1

Sahul Plate   8, 10, 12, 23
salination   76
Saxicola torquata (European stonechat), survival rates   112
scarab larvae (diet source)   53–4, 60, 121, 126
  systematic foraging and   73–5
scatter hoarders   64
Schodde, Richard, and Ian Mason   17, 18, 20, 23
scrub jays, food caching   59
Scythrops novaeholladiae (channel-billed cuckoo)   111
sea eagle, *see* white-bellied sea-eagle (*Haliaeetus leucogaster*)
sea-level changes   8, 10
seasonal variations
  dispersal patterns and   133
  food hoarding and   64, 67
secondary feathers   32, 33
security, territory establishment and   85
sedentary individuals   134
seeds, magpie diet   59
self-control   155
self-feeding   128
self-guided song learning (nestlings)   171
semantics, vocalisations   186
semi-nomadic species, cooperative behaviour   155
sensory input, brain size and   41–55
sentinels, breeding time   96–7
separation, juveniles and adults   203
sequential calls   188
Serinus canaria domestica (canary), song control system   165
serotonin, singing regulated   144
sex differences, dispersal patterns and   132
sexual dimorphism   165, 166
  breeding choice and   101
  song and   180
sexual distinction, song capacity and   42
Seychelle warbler (*Acrocephalus sechellensis*)   85
Seymour   89, 102
shade sources   76, 77
shorebirds   36, 37
  cooperative behaviour   155
short-distance dispersal, juveniles   134–5
shrikes (Laniidae)   5, 67
shrike-thrushes (Pachycephalidae)   5

silver-backed butcherbird   11
silvereyes, shared territories   81
simple calls, learning and   168–9
sittellas (Neosittidae)   5
size variations, subspecies   21, 23, 25
skeletal features   33–4
skin features   28
skinks (diet source)   59
skull   34, 35
sky-watching, eagle alarm call and   194–5
slow breeders, magpies   103
small birds, metabolic rate   57
small dragons (diet source)   59
small honeyeaters, mobbing behaviour and   95
small vertebrates (diet source)   59
snakes, as predators   129–30
snow cover, food hoarding and   64
Snowy Mountains   101–2
social behaviour, foraging behaviour and   59–60, 72–5
social environment, song learning and   131
social foraging   73
social intelligence   179–98
social learning, juveniles and   128
social organisation variations, magpie subspecies   24
social parenthood, biological parenthood and   102
social play   147–8, 218
social rules   139–58
social status, friction and   150
solicitation display   103–4
solitariness, magpies and   24
solo play behaviour   146, 147, 148, 218
song   141–4
  alarm calls and   180
  development and   167–9, 172
  functions   141
  non-directed and directed   181
  practice   167
  production   159–77
  range of   197
  reproductive capacity and   42
  territorial   170
songbirds   16
  African   5–6
  brain mass/body weight ratio   44–5, 46, 47
  cognitive mimicry   176
  cooperative behaviour   155–6
  digits of the foot   39
  learning behaviours   130–1
  origins   3, 4
  sound parameters   159–67
song learning   169–72

adults absent   131
mimicry and   172–6
sonograms
   mimicry capacity   175–6
   song development   168
sound context, mimicry   174
sound cues, scarab larvae diet   60
sound signals   182, 185–6
South American macaw, brain mass/body weight ratio   45
South Australia   3, 24, 25, 173
south-east Asia   4
southern cassowary   10
   brain size   44
Spain   112, 134
spatial attention   55
specialised feeders, endangered   213
species numbers   27–8
spiders (magpie diet)   59
spinal cord   42
splinter groups   83
spontaneous behaviours, anti-predator   129–30, 181
stable territories   80
starlings
   breeding season   103
   extractive skills   60
   vocal mimicry   172
State of the Environment 2001, 2016   76
sternum   36
stick use, food retrieval   58
storks, beak-clapping   185
striated pardalotes, nest building   106
structure, carolling   187
Struthidea cinerea (apostlebird), cooperative nest-building   156
Sturnus vulgaris (European starling), vocal mimicry   172
subharmonics, magpie song   162
submergence, continental   7, 8, 9
submissive behaviour   135–6, 151–2, 182
subspecies, Australian magpie   14–25
   G. t. dorsalis (western magpie)   10, 18, 20, 22
   G. t. eylandtensis (Top End magpie)   18, 20–1, 22, 23, 116
   G. t. hypoleuca (Tasmanian magpie)   18, 20, 22
   G. t. longirostris (black-backed magpie)   10, 18, 20, 21, 22
   G. t. papuana (New Guinea magpie)   10, 18, 20
   G. t. telonocua   20, 22, 25
   G. t. terraereginae   18, 20, 22

   G. t. tibicen   18, 20, 22, 127
   G. t. tyrannica   18, 20, 22, 23, 25, 116
sub-syringeal pressure   163
subtropical magpies   22
sulphur-crested cockatoo
   brain mass/body weight ratio   45
   mobbing behaviour   95
summer conditions, clutch size and   111
sunbathing   140
Sunda   10, 12
superb fairy-wren (*Malurus cyaneus*)   75
   communication systems   179
   cooperative nest-building   156
   cooperative structures   135
   genetic assertiveness   102
   shared territories   81
   song function   167–8
superfamilies, songbirds   4, 6
survival rates, offspring   111–14, 127
swearing capacity   176–7
swooping behaviour   199, 206–11
Sydney Morning Herald   1
syllables, magpie song   163
synchronous hatching   118
syndactyl foot   39
syntax, vocalisations   186
syringeal muscles   159, 160–1, 167, 168, 174
syrinx   159, 160–4, 166, 168, 185
systematic foraging   73–5

Tachycineta bicolor (tree swallow), non-hierarchical feeding   119
Taeniopygia guttata (zebra finch), *see* zebra finch
tail feathers (rectrices)   33
   courtship and   103
   loss of   217
tarsometatarsus   37
Tasmania   24, 25, 173
Tasmanian magpie (*Gymnorhina tibicen hypoleuca*)   18, 20, 22
Tasmanian native-hen, crouching behaviour   151
Taveuni   195–6, 222
tawny frogmouth
   asynchronous hatching   118, 119
   cuddling   144
taxidermic predator models experiments   51–2
taxonomy, magpies   17, 23–6
teamwork, territorial defence and   86, 95–6
temperature range, nestlings   122
Tephrodornithidae (woodshrikes and allies)   5
territorial survey, individual humans and   205

territory
  bonded adult magpie pairs and   66, 101
  corridors   91–2
  establishment of   84–6
  evictions from   80–1
  food caching and   65
  food foraging efficiency and   72–3
  human territory and   82, 199
  leaving of   127
  management of   79–98
  nest building and   105
  play fighting and   148–50
  quality of   85–6
  size   86
  shifts   134
  song sharing   170
  status classification   83–4
  vocal mimicry and   173–4
territory defence
  carolling/songs and   86, 141, 167, 181, 186, 187–9
  cooperative   156–7
  courtship and   103
territory warning, Brisbane City Council   208
Tertiary fossils   2, 10, 12
testosterone levels, defence skills and   96
Thamnophilus atrinucha (western slaty antshrike), survival rates   113
theft, conflict over   151
thermoregulation   76
thighbone   37
thoracic air sac   163
thoracic vertebrae   36
thornbills (*Acanthiza* ssp.)
  mobbing behaviour and   95
  shared territories   81
  vertical territory   63
tibiotarsus   37
time constraints, learning behaviour   71
tit berrypecker (Paramythiidae)   5
tool use   61
Top End   13, 134
Top End magpies (*Gymnorhina tibicen eylandtensis*)   18, 20–1, 22, 23
top predators, dangers from   203–4
Torresian crow (*Corvus orru*)
  brain mass/body weight ratio   45
  diet   61
  origin   23
Townsville   24
Toxostoma rufum (brown thrasher), two-voice vocalisation   163
trachea   159, 162, 164
tranchobronchial syrinx   161

transient birds, magpie territoriality and   174
translocation, Pacific region   214
tree of life, avian   4
tree swallow (*Tachycineta bicolor*), non-hierarchical feeding   119
trillers (Campephagidae)   5
tropical magpies   22
truce behaviour
  humans and magpies   206–7
  territory defence   90–1
tubers (magpie diet)   59
twig weaving (nests)   105
twine injuries   202
two-voiced vocalisation   161, 163

ulna   32
underground food sources   58
  see also extractive foraging skills
uninterrupted singing, mechanism   162
University of New England   35
unpalatable food, dunking and   67
unprotected nests   108
unstable territories   80
uropygial gland (preening)   28

Vanellus miles (masked lapwing)
  hatchling incubation   117
  mobbing behaviour and   95
vangas (Vangidae)   4, 5, 6, 8
  African links   11
Varanus varius (monitor lizard)   93, 109
venomous creatures   203
vertebrae   35–6
vertical feeding zones   63
vervet monkeys, water access defence and   90–1
Victoria   3, 89, 95, 102, 108, 156, 173
vigilance behaviours   130, 182
  territory establishment and   84
Vireo atricapilla (black-capped vireo)   86
Vireonidae (vireos)   5
visibility, feathers and   32
visible obstructions, vocalisation quality   182
vision   53–4
  binocular   128, 197
  lateralised   53–4
visual cueing, food items and   71
visual processing, contralateral hemispheric lateralisation   48, 49, 50
visual reaction times, learning behaviour   71
vocalisations   159–77
  anti-predator   129, 130
  categories   181
  defence strategy   87, 88, 89

learning of 41, 130–1
mimicry 172–6
song production 167
see also song; songbirds; song learning
voice recognition 185–6

waders 37, 46
walk-foraging 53, 63
walnuts, magpie diet 59
Walpa Gorge 221
warbling
carolling and 187, 188
duration 142, 143
warning signals 185, 206, 209
water competition 90–1
water drinking, hatchlings 119
water engagement 141
waterway diversion 79
wattle-eyes (Platysteiridae) 5
weaning 129
weather patterns, feeding patterns and 75–8
wedge-tailed eagle (*Aquila audax*) 82
defence against 93, 157
magpie perception of 51, 52
weight increase, nestlings 124
Western Australia 2, 3, 6, 7, 14, 17, 22, 24, 25, 77, 81, 103, 173
magpie foraging 75
western magpie (*Gymnorhina tibicen dorsalis*) 18, 20, 22
western slaty antshrike (*Thamnophilus atrinucha*), survival rates 113
West MacDonnell Ranges 77
westward movement, magpies subspecies 23
whipbirds (Psophodidae) 5
whistlers (Pachycephalidae) 5
white-backed magpies (WBM) 17–18, 20, 23
defence 87
sexual dimorphism 101
white-bellied sea eagle (*Haliaeetus leucogaster*)
defence against 157, 220
white-breasted woodswallow (*Artamus leucorynchus*) 34
white-crowned sparrow (*Zonotrichia leucophrys*), song activity 172

whiteheads (Mohouidae) 5
white plumage 19, 31–2
white-winged chough (*Corcorax melanorhamphos*) 5, 61
cooperative nest-building 156
cooperative structures 135
mobbing behaviour 95
non-hierarchical feeding 119
wild dogs, teamwork 96
Wildlife Information and Rescue Services (WIRES), attrition statistics 200, 201, 203
willy wagtails, mobbing behaviour 96
wind direction, nest sites and 108–9
window of return, fledglings 135–6
wing
flapping 185
swishing 156
venting 76, 77
wolves, teamwork 96
wood pigeons, wing flapping 185
woodshrikes (Tephrodornithidae) 5
woodswallows (Artamidae) 4, 5, 11, 13
brain mass 45
camouflage 32
hopping 37
upper canopy feeding 63
Working List of Australian Birds (Birdlife Australia) 14, 17
World Bird List (IOC) 14, 17
Wulst area 42–3, 51

zebra finch (*Taeniopygia guttata*)
body weight/neuronal density ratio 47
brain lateralisation 50
brain size 43, 46
duetting calls 188
mating compatibility 101, 112
song control system 165, 166
song learning 131, 169, 180
trachea 164
water sources 91
Zonotrichia leucophrys (white-crowned sparrow), song activity 172